ROOFS AND ROOFING

Design and Specification Handbook

ROOFS AND
ROOFING
Design and Specification Handbook

Published by
Whittles Publishing,
Roseleigh House, Latheronwheel,
Caithness, KW5 6DW, UK
Tel. and Fax. 05934-240

ISBN 1-870325-11-7

British Library Cataloguing in Publication Data

Coates, David T.
  Roofs and Roofing : Design and
  Specification Handbook
  I. Title
  690

Text layout and Cover Design by
Smith & Paul Associates,
Balgonie House, Balgonie Estate,
Paisley, PA2 9LN
Tel: 041-884-2212 Fax: 041-884-7802

Printed by Bell & Bain Ltd

ROOFS AND ROOFING

Design and Specification Handbook

David T. Coates

MCIB, FFB, FIoR, MSE, PEng

Roofing Consultant

This book is dedicated to
MUM and DAD
with thanks for their love, guidance and encouragement

CONTENTS

ACKNOWLEDGEMENTS

The writing of this book, which covers such a wide range of subject matter, would have been a far more laborious task but for the help and advice of many experts within the roofing industry.

In particular I would like to record my thanks to Ken Gunn for his collaboration in those sections dealing with tiles and slates, and to Brian Caldwell for his corresponding involvement in respect of felts and membranes. I am also extremely grateful to Martyn Duncan for his work in producing almost 100 drawings from my original rough sketches.

I also thank the Avoncroft Museum of Buildings for allowing me free access to an extensive library, and the management of the London Planetarium for providing me with copies of archive material on the construction of their domed roof.

Finally, I would like to mention all those colleagues and associates who have helped and encouraged me by their advice, enthusiasm and technical comments. The list of names would be far too long for this page, and their combined contribution was enormous. I thank them all.

DTC

FOREWORD

The roof is, arguably, the most important element of a building. It is the roof which provides the primary protection against rain, wind and snow. In the case of single-storey buildings, the structure is intended to support the roof, and the perimeter walls simply close the space between the roof and the ground (almost all roofing materials could be used for vertical walls).

It sometimes seems that the building and construction industry has failed to recognize priorities. There are numerous books on structural design covering steel frames, reinforced concrete, timber, load-bearing brickwork, raft foundations and piles; there are many volumes on each topic. As new structural theories are evolved, new books are written. These may show theoretical justifications for marginal reductions in safety factors, or second-order effects in the development of design stresses.

I have long been puzzled by the dearth of books on roofing. After all, without a roof there would be little need for a building structure. Furthermore, it is the roof which provides protection against the elements. Because of the roof, the structure is kept dry and maintained at a fairly even temperature and it is the roof which experiences the extremes of temperature between winter and summer, drenching by rain, ultraviolet radiation and atmospheric pollution. When the roof ceases to resist these effects, the enterprise within the building is damaged or stifled.

In recent times, roofing has increased in complexity. The use of thermal insulation has grown dramatically, and has brought about the introduction of vapour checks and breather membranes. New products have been developed in response to market pressures, and new construction methods have been evolved.

These changes have resulted in a significant increase in the cost of the roof as a proportion of the total cost of the building. At the same time, the proliferation of roofing products has served to reduce the depth of knowledge that a designer or fixer can achieve in any single product or system. It is against this background that my book was conceived and written.

The subject is far too broad to be covered in depth in a single volume. I have identified what I consider to be the eighteen essential topics, and have devoted a chapter to each. I have attempted to state the basic facts on each topic, and suggested further sources of information. There is a further fund of knowledge available from trade literature, and I hope that my book will provide sufficient information to allow the reader to understand and assess the claims which manufacturers are apt to make for their products.

Roofing cannot take place in secret and it is not confined to remote or inaccessible cities. Roofing is happening all around us, every day. A student of roofing has ample opportunity to see many products and their application, and to ask questions of their users and fixers.

Roofing is an important subject in its own right. It is to be hoped that the efforts of such bodies as the Institute of Roofing and the National Federation of Roofing Contractors will help to bring about a wider recognition of our industry.

EXPERTO CREDE

The Institute of Roofing

PROFESSIONAL STATUS IN
THE ROOFING INDUSTRY

by Graham Bateman, FIoR, Director, Institute of Roofing

The contribution made by the roof in the putting together of any building project is all-embracing, and it is largely under-valued. The roof is vital to the purpose demanded of all buildings, i.e. their exterior and interior well-being. It can do much as a medium for the designer's aesthetic and technical skills, and for all this, although it can be the subject of merciless economies, it is rarely an element of major consequence in the overall compilation of costs.

It has also to be said that in its simpler forms it lends itself to exploitation by what we have come to describe as 'cowboys', those who have no particular allegiance, or no allegiance at all, to the good name of the industry. This element in the industry has long been a target for its detractors and those who have taken on the duty of protecting the interests of the public.

It was this that motivated the setting up of the Institute of Roofing in the early 1980s. Its group of 400 founder members came largely from executives drawn from major firms of roofing contractors and some of the principal manufacturers. Although it had the blessing and some assistance in its initial funding from the National Federation of Roofing Contractors, the Flat Roofing Contractors and Advisory Board, and the Mastic Asphalt Council and Employers Federation, all its members are recognised in their private capacity only. Their brief was to devise a code of professional standards in the management of roofing contracts, and provide professional qualifications which would serve as an incentive to recruitment and promotional advantage within the industry.

Today it is now fully independent, with a membership which has suffered losses during the recession but is still within the measure of 1200 in the grades of Fellows, Members, Associates, Licentiates and Affiliates. They come not only from roofing contractors but also from manufacturers, all committed to the development of a professional and technically competent roofing industry. They are served not only from headquarters but by a very active regional organisation which promotes seminars and works visits and occasional social functions. The annual general meeting has become something of an event in the industry, attended by many leading personalities. It incorporates a lively discussion session and an address by a prominent guest speaker.

The Institute's eventual objective is to see admission to membership restricted to the examination room, but while it remains a comparatively young organisation with many potential members who, because of their seniority, could not be expected to take a formal examination, there is a strictly controlled facility for direct entry by those of sufficient status and experience to qualify for admission as Fellows or Members.

The Institute is now playing an important part in the devising of National Vocational Qualifications for careers in the industry's management and technician sector.

As competition in the single market becomes more widespread, with increasing insistence upon quality assurance as a condition for inclusion in the tender lists of public authorities, the Institute is pressing upon employers the advantages to be gained by encouraging staff to qualify for IoR membership as evidence of professional competence when seeking work in the upper range of contract values.

Further information is available from:

> Institute of Roofing
> 24 Weymouth Street
> LONDON, W1N 3FA

NATIONAL FEDERATION OF ROOFING CONTRACTORS

The National Federation of Roofing Contractors Ltd (NFRC) celebrates its Centenary year in 1993 having been constituted in its present form in 1943 by the amalgamation of roofing craft associations dating from the 19th century. Its main role is to ensure that high standards of workmanship, specifications and materials are applied to all roofing activities to achieve high quality roofs.

The Federation's membership consists of Associate manufacturing companies and Trade members - the contractors. The latter are carefully vetted prior to acceptance. Thereafter in order to become Registered members of the NFRC they must undertake to abide by the Federation's Code of Practices. Each is issued with a Registered number and then is allowed to offer the unique Copartnership scheme - an insurance package which covers both materials and workmanship and offers a ten year warranty on roofs fixed in accordance with British Standards both for domestic and corporate clients. In conjunction with this scheme the Federation has a disputes procedure which specifiers and clients can call upon, without prejudice to their legal rights, should they consider that the workmanship of a Federation trade member is unsatisfactory.

The Federation supplies its membership with a wide range of technical information, protects its members' interests in the harmonising of roofing standards in Europe, lobbies government on new legislation affecting Roofing Contractors as well as regularly advising and updating them on new methods, Quality Assurance, products and services, changes in Health and Safety legislation, prices, wages, conditions of employment and contracts etc.

Considerable importance is attached to training and the Federation works closely with the CITB Regional Roofing Training Groups, technical colleges, skill training centres and schools to encourage people into the Roofing Industry and to achieve higher skills. A number of scholarships and awards are sponsored by the Federation. The NFRC also produces regular technical publications including the journal ROOFING and an Annual Report and an Organisation Directory which identifies "Who's Who" in the Roofing Industry.

If you are interested in more information contact:-

NFRC
24, Weymouth Street
LONDON, W1N 3FA
Tel No: 071 436 0387
Fax No: 071 637 5215

HISTORY

Plate 1.1   A thatched roof with dormer windows and a porch canopy.

A subject can only be understood and appreciated by including its history in its study. Past influences such as knowledge gained by experience, scientific discoveries or changing fashions have shaped what we see today.

It is only possible to understand the reasons for the adoption of specific materials, roof shapes, structural support systems and multilayer combinations when these are seen in the context of their development. The roofing systems used today have been improved by practical experience; the improvements may be in cost, durability, ease of fixing, weathertightness, or many other aspects of performance. This chapter seeks to describe the changes, fashions and influences which together have brought the British roofing industry to its present position.

Once having made a decision to consider the history of roofing, a further decision is required: what is the starting point? Any choice here can only be arbitrary, but the Roman occupation seems a logical point, as it is fairly certain that there was little building of significance in Britain prior to the arrival of the Romans.

The Romans brought a knowledge of several aspects of building technology with them, and this affected their roads, bridges, fortifications and buildings. In our case, we are interested in the Roman approach to roofing.

There is no doubt that the Romans used slate and other natural stones which could be split into thin pieces; they were ready to exploit any convenient local resource. They were also skilled in the manufacture of tiles, which like bricks, were formed

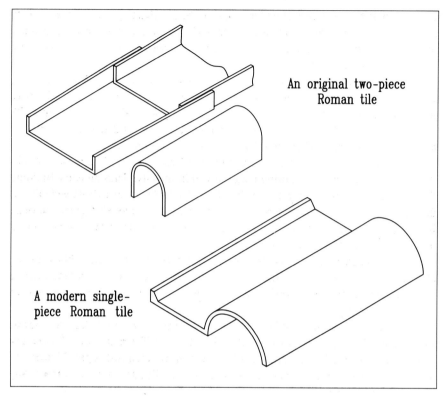

An original two-piece
Roman tile

A modern single-
piece Roman tile

Figure 1.1   Some 'modern' tile designs have a long history.

from local clay and fired in kilns. Tiles are produced in Italy today to essentially the same pattern as those produced by the Romans 2000 years ago. In fact, one of our current tile shapes, described as 'Roman' has a similar appearance to the old Roman design, but is formed in a single piece rather than the two pieces required by the Romans.

After the Romans left Britain, many skills and crafts were lost; there is no evidence of tiling or slating for many centuries. Roofs in Britain were covered with the most basic materials, such as thatch, bracken and turf. These roofs could not possibly have survived for long periods, so the evidence of their existence must be sought in contemporary records.

Thatch is mentioned in the writings of the Venerable Bede, who died in AD 735. It is beyond doubt that an economy based on agriculture, and in which cereal crops were of high importance, would have had a surfeit of straw. There can have been little incentive to search for other materials.

Thatch is a lightweight material, and this has advantages in its handling and fixing, as well as in the design of the supporting structure. A thatched roof must have very steep slopes if it is to keep out the weather, but this was not a problem when buildings were small. The material provides good thermal insulation, making for a comfortable dwelling, warm in winter and cool in summer.

As people began to congregate in cities, so the disadvantages of thatch began to appear. One of these was simply the problem of transporting straw from the country to the city in sufficient quantity. A material is no longer free when resources are required for its transport.

However, the greatest disadvantage of thatch was its flammability, and this eventually ended its use within cities. The records show that there were major fires in London in 1077, 1087, 1135 and 1161, Canterbury in 1161, Winchester in 1180, Glastonbury in 1184, Chichester in 1187 and Worcester in 1202. It is easy to imagine how a fire would spread between thatched roofs in close proximity.

In London the problem was addressed by the Ordinance of 1212, which prohibited the use of thatch as a roofing material within the city. Thus an early Building Regulation existed three years before the Magna Carta! Other cities followed suit, but the pace was very slow by modern standards; for example, it took a further 400 years before thatch was prohibited in Cambridge. Of course, thatch continued to be popular in the country, and remains so to this day.

Clearly thatch could not have been prohibited unless there were readily available alternatives, and these alternatives had to offer some advantages. In fact, there is evidence that several other roof-covering materials were available at the beginning of the 13th century.

The Bayeux Tapestry, produced in France between 1077 and 1082, includes an illustration of a roof covered with oval timber shingles. There are also 13th-century references to boarded roofs, presumably planks with overlapped edges. Timber was a common enough material, and there was already a tradition in carpentry for building ships as well as houses.

It is also well documented that timber prices rose dramatically in the 14th-century, and this must have stifled the growth of shingles and boards as roofing materials.

The use of slate was revived in the 12th century, but only in those areas where it was naturally available — there was no transport system for the mass distribution of building materials. However, slate was widely used in Cornwall, Devon, Leicestershire and the Lake District. Other natural stones were used in similar ways, for example limestone in the Cotswolds and sandstone in both Northumberland and Hereford. Each area made the best use it could of its own natural resources.

Slate can be sliced thinly, but some other stone materials must be used in greater thicknesses, and this results in heavy roof coverings. Heavy materials need more substantial structural support. The use of irregular pieces of stone resulted in a need for steep roof slopes in order to maintain weathertight details. So the change from thatch to slates called for stronger rafters, but left the roof shape unchanged.

It is not known exactly when the craft of tile-making was rediscovered, but certainly tiles were in regular use in the 12th century in Britain. They were made from local clay and, like slate, were too heavy to transport over significant distances. Tiles were formed to regular shapes, and incorporated beneficial features to aid fixing or jointing; they were also able to provide weather protection at slightly lower pitches than was possible for slates. They were probably a little heavier than slate, but lighter than most other natural stones.

Tiles and slates offered the enormous advantage of being non-combustible, and this must have been seen as a great blessing in cities, where so many disastrous fires had occurred. They were also far more durable than thatch or timber, and this must have had economic consequences for the purchaser and the builder.

It seems likely that thatching, in the earliest times, was carried out by people on their own dwellings. It was just another skill like ploughing, sowing and harvesting. The advent of slates and tiles almost certainly resulted in specialization by dedicated roofers who learned the skills and techniques, and applied them as a craft or trade. This seems to be borne out by reference to modern surnames; 'Slaters' and 'Tylers' are more numerous than 'Thatchers'.

The pace of development was very gentle and it is not known exactly when metal began to gain some popularity in specific applications. At first, the metal was lead, but other possibilities were eventually tried. Lead has the great advantage that it is soft, and easily formed or cut in situ. It can be permanently joined by soldering without the need for excessive heat, and the joint is totally weathertight. It is extremely durable, and makes long maintenance-free life a possibility.

Its two greatest disadvantages are its weight and its lack of mechanical strength. At a time when all building materials were lifted, moved and fixed by physical effort, the mass of lead cannot have been welcome. Lead was used as a durable, high-quality weather protection, but it was not capable of spanning between purlins, or other supports. The first stage of making a lead roof was to lay a continuously boarded platform to receive the lead.

Soldered joints made relatively shallow roof slopes possible. This was of great importance in the design of churches and cathedrals; some of these had wide span roofs, and if these had required steep slopes they would have risen to unacceptably high ridges.

Copper also gained a very gradual acceptance. Its cost was too great for it ever to become a common form of roof covering, but it achieved a limited following for its attractive appearance, and its extreme durability. It is recorded that Sir Christopher Wren originally specified the use of copper cladding for the dome of St Paul's Cathedral; however, on learning that the cost would be around £3000, the government refused him the money, and insisted on lead covering at a cost of £2000!

Aluminium was not isolated in metallic form until 1826. Even then, it was very expensive, and it was nearly another hundred years before it was available in commercial quantities for roofing. One of the oldest examples of aluminium roofing is the church of San Gioacchino in Rome. This roof was built in 1897 and, after almost 100 years, remains in excellent condition.

Zinc, galvanized steel and stainless steel are also used occasionally. Continuously supported metal has never been widely used; it has been seen as a prestige product for specialist applications.

The industrial revolution affected the roofing industry in several ways. Improved transport, first by way of the canals and later the railways, made it economically viable to move bulky, heavy materials over relatively large distances. In particular, this newly available transport boosted the use of slate as a roofing material. Slate of superb quality was available from several areas of Britain but these areas had a common feature — they were remote from the major cities. The slate quarries of Wales, the Lake District and Cornwall could never have developed without the means to transport the product to the customer.

The industrial revolution created a rapid increase in the amount of building. New homes were needed to house the thousands of workers who were leaving the countryside to find work in the cities. There was a further demand for roofing for the new factories. The slate industry expanded to meet these growing needs, and for a while slate enjoyed the major share of the market.

Where individual taste is involved, it is not possible for a single product to dominate a market completely. As more and more roofs displayed a smooth blue-grey appearance, so there was an increasing demand for something obviously different. The answer was found in tiles, which could be formed into special shapes, and which could provide welcome alternative colours.

The new transport systems were able to cope with moving the tiles, and modern factory production methods assisted in achieving efficient manufacture.

As industrial processes changed, there was a corresponding change in the design of factories. Most early factories were essentially similar to houses; when factories grew larger they were rather like large houses. The traditional construction used steep-pitched trusses over relatively small spans; wider buildings had several such

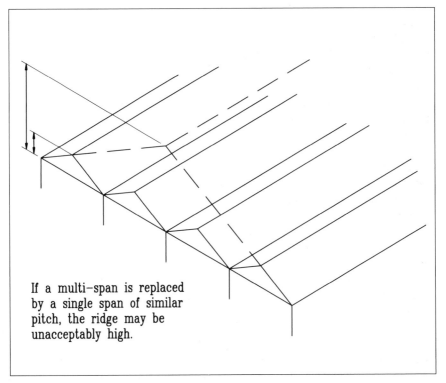

If a multi-span is replaced by a single span of similar pitch, the ridge may be unacceptably high.

Figure 1.2   The relationship between span, pitch and height.

trussed roofs side by side. This form of roofing requires structural supports under the valleys, but it also requires runs of gutters between adjoining roofs, and this further demands rainwater pipes to drain the gutters. All of this serves to reduce the amount of clear space within the building.

The factories not only needed large spaces for the new machines, they needed the flexibility to accommodate even newer machines as they became available. To achieve this they needed wide-span roofs without internal drainage.

There were two major obstacles to the development of wide-span factory roofs. First, roof coverings such as slates and tiles were heavy; a wide span involved supporting a heavy dead load due to the self weight of the roof. This represented unacceptable expense. Second, roof pitches were necessarily steep in order to achieve weathertight construction; the combination of wide span and steep pitch can only lead to grotesquely high ridges, and this was also unacceptable.

The great breakthrough came in 1913 when Turner Bros opened their Manchester factory for the production of asbestos cement roofing sheets. This corrugated roofing was cheap and light, could be fixed at lower pitches, and rapidly revolutionized factory roof design.

The low weight of this product was partly because the material's basic mass was low, and partly because the sheets were very large by comparison with slates. The areas of overlaps were much reduced, so most of the roof was covered with a single skin of asbestos rather than a multiple thickness of slate. In the 'as fixed' condition, asbestos cement was considerably less than half the weight of slate.

The material made lower pitches possible by virtue of its profiled shape which simplified side laps, and its availability in long lengths which reduced the number of end laps. This product so exactly fulfilled the needs of industrial roofing that for over 50 years asbestos cement ruled supreme.

Factories also needed light to enable the occupants to see their work. This light was at first provided by way of glazed openings in the roof; free lighting was to be preferred whenever possible. Various systems were developed, usually consisting of glass supported in specially designed bars. So many systems were developed and patented that users became confused between the different products; it became common practice to refer to all such systems as patent glazing, and this generic term continues in use to this day.

Initially, most factories generated heat as part of their production processes. Typically this could have been in steam boilers used to generate power, or in furnaces for metalworking. But many factories had surplus heat, and had no incentive to conserve it. Even factories which housed cold operations made no great expenditure on heat, as the rules governing conditions in factories did not require high temperatures for worker comfort. In the very few factories where adequate heating was provided, the costs were not high, as energy was cheap. The concept of energy efficiency was not seriously considered until the late 1950s.

Industrial action in the mining and electricity industries began to cause power cuts and shortages. At the same time, the OPEC organization raised the price of oil, energy prices increased dramatically, and this created an incentive to conserve heat. The use of thermal insulation expanded rapidly in both commercial and domestic

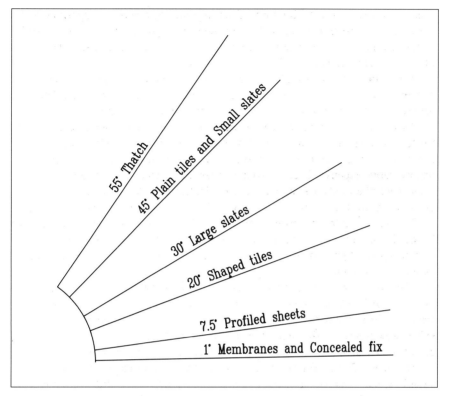

Figure 1.3 Typical minimum pitches for various roofing products.

roofing. New insulation materials appeared on the market, and some older ones were used in greater quantities.

Rapid changes usually involve some unexpected problems, a side-effect of using more thermal insulation was a major increase in the incidence of condensation, which contributed to corrosion and decay. The perceived solution was the introduction of additional layers in the roof construction, the vapour barrier and the breather membrane.

Insulation also increased the thickness of roof construction, and most insulants were easily crushed. New fasteners and fixing techniques were thus required.

Simultaneously, other changes were taking place. Coated steel and aluminium had started to acquire a significant share of the market for wall cladding, they were aesthetically attractive and were more robust than asbestos cement. During the 1970s, coloured metal started to take a little of the industrial market. Then, early in the 1980s, health hazards were identified in respect of asbestos, and the asbestos cement roofing industry declined rapidly. In the industrial roofing sector the loss of market share, by asbestos cement, was spectacular — from over 80% to under 15% in ten years. The beneficiaries were steel and aluminium, with steel taking by far the greater share.

The use of natural lighting, in industrial buildings, had promoted the development of profiled plastic sheets which could be used in conjunction with profiled roof sheets. As energy conservation became popular, so double-skin rooflights were produced. These were not particularly efficient in conserving heat, and created great difficulties when used in conjunction with vapour barriers and breather membranes.

When industrial roofs consisted of single-skin profiled sheets, the internal appearance was quite simply governed by the appearance of the back of the roof sheets. With the use of insulation this changed; quilt insulation required the use of a lining sheet to support the insulation.The lining sheet could be selected for its appearance. Alternatively, some insulants were available as boards which could span between purlins; these boards were given decorative finishes on their underside.

The use of roofing felts was established over a number of years. The typical felt had fibre reinforcement, and was impregnated with bitumen. During the 1960s and 1970s there was a trend towards the design of flat-roofed buildings. This was encouraged by the planners, and supported by architects. Multi-layer felt roofing was frequently specified for roofs with virtually no fall, and with a multitude of openings.

Many unsatisfactory roofs were built during this period, due partly to poor materials, partly to poor workmanship, and partly to poor design details. Felt roofs acquired a bad reputation which was not easy to overcome. In due course, better membranes were developed, and fixing methods were also improved. Flat roofs could now be as durable as other types of industrial roofing.

Another version of flat roofing, the concealed fix system, appeared. Concealed fix roofing, as its name implies, has the fixings concealed from the weather in such a way that the screw holes cannot become leak points. The weather skin consists of interlocking metal panels, and it is a feature of the systems that these panels are supplied in great lengths (sheet lengths of over 40m have been transported to site in Britain!). The absence of end laps, secure interlocked side laps, and impenetrable fixing points, allow the use of these systems on very shallow slopes.

Developments in domestic roofing have been less dramatic. Thermal insulation has been introduced, just as in industrial roofing, but the main change has been from clay tiles to concrete tiles. The first Redland factory for concrete tiles was opened in 1919 and the first Marley factory was opened five years later.

The new concrete tiles were a little heavier than the traditional clay tiles, and did not have such attractive or permanent colours. However, they were competitively priced and extremely durable. Concrete tiles penetrated the market at a remarkable rate. Today they enjoy a major share of the domestic roofing market.

This change in material has not had any great influence on construction methods, and domestic insulation is often laid at loft floor level, rather than being incorporated in the roof slopes. Because changes have been small, skills have been developed over several generations of designers and craftsmen.

By comparison, the changes in industrial roofing have been both varied and numerous. Designers and craftsmen have not had the opportunity to learn from mistakes, or to refine successful techniques; no two roofs are alike, and materials appear and disappear at bewildering speed. Users must take care to study the product literature, and to take advantage of any training offered by materials manufacturers.

During the past 50 years, a greater understanding of wind loading has been achieved. It has been shown that wind uplift forces become larger as roof pitches are reduced. This points to another difference between domestic and industrial roofs: in industrial roofing, the pitch has been progressively reduced, while its self weight has also been reduced. Consequently, the uplift forces have increased, and the self weight to counteract the forces has been reduced, so that fasteners have assumed greater importance. By comparison, domestic roofing has not changed in pitch, so the wind loads are unchanged, but the self weight of the roofs has increased.

Although new materials and methods are developed continually, any traditional systems which have proved themselves in service have been retained and improved. For example, thatch is widely used on country cottages; modern thatch is usually made using Norfolk reed, as this provides the best durability. A Norfolk reed roof should give at least 50 years of satisfactory service, and 100 years is possible.

Similarly, fully supported metal roofs are sometimes used in applications calling for exceptional durability. Churches and cathedrals are intended to survive far longer than houses and factories; fully supported lead, copper or aluminium can provide long-lasting, maintenance-free roofing, even at very shallow pitches. Modern manufacturing methods allow metals to be produced in greater purity, as improved alloys or with more uniform properties.

Examples of roofing, past and present, are very easy to find. Every town has old houses and new estates, old halls and modern offices, and factories of various ages. Villages have cottages built from local materials, and farm buildings in the style of the area. It is not often necessary to make extensive searches for roof types. A student of roofing will find many interesting examples within a few miles of his home, and will soon learn to observe unusual local features in roofs he encounters during his travels.

STANDARDS
The Building Regulations

FURTHER READING
The Pattern of English Building - Alec Clifton Taylor - 1972
The English Medieval House - Margaret Wood - 1966

ROOFING CONFIGURATION

2

Plate 2.1 Multi-span barrel vault roofs in profiled metal.
(By courtesy of Metal Constructions Ltd.)

Almost every specialist subject has its own terms and definitions, and roofing is no exception. The writers of manuals, standards and manufacturer's literature usually assume that the reader will know the basic jargon, and it is therefore important that any special terms are clearly understood.

The most fundamental property of a roof is its *pitch*. The pitch is the angle between the line of the roof and the horizontal; a small pitch means that the roof is almost flat. A mathematician or engineer would probably prefer to use the word 'slope'or 'gradient', but this would lead to confusion in roofing. A roof *slope* describes any plane of roofing where the pitch is constant.

Some continental manufacturers express the pitch as a percentage, they state the vertical rise for a horizontal distance of 100 units. Thus a $10°$ pitch is 17.5%, and a $30°$ pitch is 58%.

It is possible to have a roof with zero pitch, but such an arrangement is unusual. When rain falls on a roof it must be drained away to the gutters, and the greater the pitch, the more efficient the drainage. A completely flat roof would almost inevitably have standing water because structures deflect and settle, creating low spots.

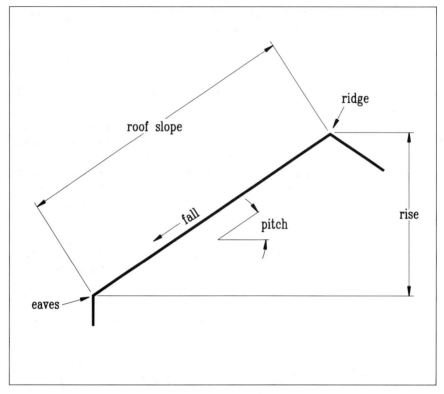

Figure 2.1   Definitions of the major geometrical properties of roofs.

Standing water (usually referred to as *ponding*) should be avoided wherever possible, as it is more likely to cause leaks, or corrosion, than running water, which has a more transitory nature.

The highest point of a roof slope is usually called the *ridge*, and the lowest point is usually called the *eaves*.

The word *fall* is sometimes used to describe the 'downhill' direction. This can be useful on a roof plan where the direction of drainage would otherwise be unclear. It is also helpful in descriptions, e.g.:'The roof falls from Grid Line E to the West boundary'.

The *rise*, on the other hand, has a much more specific meaning. It is the vertical distance between the top and bottom of a slope, i.e. between the eaves and the ridge. These terms, and the relationship between them, are illustrated in Figure 2.1. They should be committed to memory, as they are essential in understanding the remainder of this chapter and in discussing roofing matters in general.

By far the most common roof shape is the *symmetric* roof with a central ridge. This is the favoured arrangement for houses, factories, barns, sheds, and many other buildings. It usually provides the simplest details, and the minimum work in design and construction.

It is possible to locate the ridge at one side of the roof, and this configuration is described as a *monopitch*. There are several reasons why a monopitch design may be adopted. The roof may be very narrow, and a ridge could then be an unnecessary complication. There may be some difficulty in providing drains at one side of the building; in such cases it is a clear advantage to make the roof drain to the opposite side. The latter case might apply in the design of a grandstand, where one side must be completely open.

There are also an infinite number of possibilities for *asymmetric* roofs. These could have a central ridge but unequal pitches, or an offset ridge with either equal or unequal pitches. Such roof forms are sometimes used to provide an aesthetic effect, but are more often the result of meeting the design criteria. For example, if a house is to have wonderful views in only one direction, it may be designed as double-storey at the front, and single-storey at the rear; this would be a good application for an offset ridge.

One special form of asymmetric roof is the *North Light*. This roof shape enjoyed a period of popularity with designers of factories and other commercial buildings. The name derived from the fact that the roof was built with its steep slope facing north, and this slope was glazed. This gave the benefit of uniform lighting, without

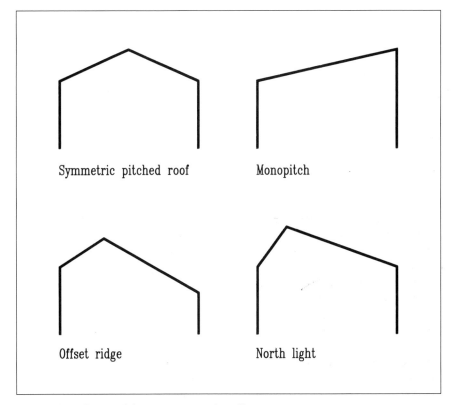

Symmetric pitched roof   Monopitch

Offset ridge   North light

Figure 2.2   Some of the common roof profiles.

any uncomfortable heating from strong sunlight in summer. It was relatively simple to install opening lights in the steep, glazed slope; this provided a means of ventilation. The arrangement is now little used because the glazed areas can allow large heat losses.

A fairly common symmetric variant is the *mansard* roof. A mansard has two distinct slopes on either side of the ridge. The slope which starts from the eaves has a large pitch; from the top of this slope there is a second slope to the ridge. It is a characteristic of mansards that the second slope is of smaller pitch than the first.

Mansards are often employed to increase the amount of usable roof space. The steep pitch, from the eaves, is used to create a rise to about head height; thereafter the shallow pitch is used to prevent the roof from becoming too high.

Mansards are sometimes used for aesthetic reasons. For example, a multi-storey building in a street cannot be viewed from a distance; this can prevent its roof from being seen, and this may detract from the overall appearance of the building. A very steep mansard slope can overcome this problem.

It is also possible to use highly decorative, but expensive, materials on the visible slopes of the mansard, and to economize on the upper slopes. This method can be particularly useful when large new buildings must blend with older traditional buildings.

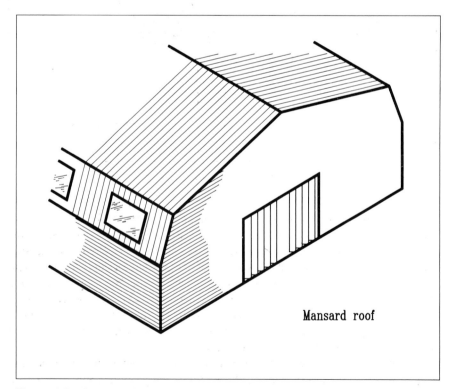

Mansard roof

Figure 2.3   A mansard roof has two distinct pitches: a steep pitch near the eaves, and a shallow pitch near the ridge.

There are several sound reasons why windows should be fixed in a vertical plane. This can be arranged, in a steep mansard slope, by simply tapering the window jamb, whereas in a conventional roof slope it is only possible to use skylights.

Occasionally it is imperative that a roof drains to the centre. Canopies over petrol pumps can provide an example of this requirement; the space around the canopy must be completely clear, and the supports (and rainwater pipes) must be central, in the same line as the petrol pumps. This arrangement is described as a *butterfly* roof. The shape is avoided whenever possible as it is usually preferable to drain to the outside of a building.

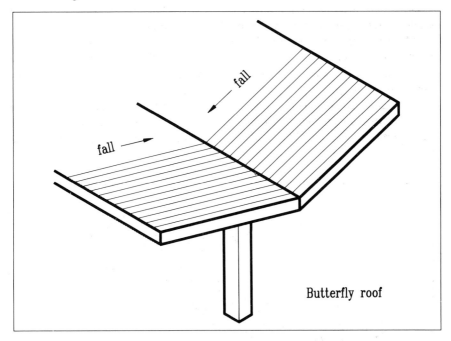

Butterfly roof

Figure 2.4 A butterfly roof is lower in the middle than at the edges.

All the foregoing roof shapes were made up from straight lines, but it is also possible to have curved shapes. Various types of arches may be encountered, but the most common is the *barrel vault*. This roof takes the form of a circular arc in cross section. It must be appreciated that such a roof does not have a single specific pitch; the pitch changes constantly around the curve, and is zero at the top of the arc. There is no ridge. Not all materials are suitable for forming barrel vaults; an appropriate material must either be sufficiently flexible to bend around the curve, or be applied in such small pieces as to effectively follow the curve. The materials must be capable of operating at zero pitch, at the top of the arc.

Curved roofs have obvious aesthetic advantages, but there are sometimes practical advantages as well. During World War II many aircraft hangars were built with curved roofs. It was shown to be simpler to camouflage a roof which did not have sharp corners.

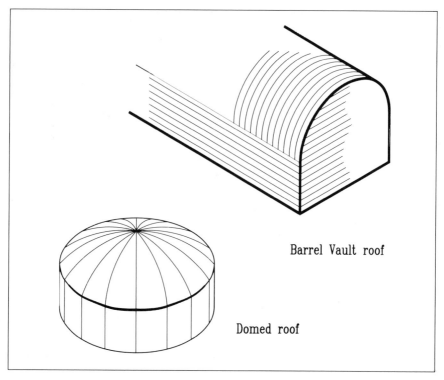

Barrel Vault roof

Domed roof

Figure 2.5   The two most common forms of curved roof.

The other form of curved roof is the *dome*. This is almost always used for aesthetic effect, and calls for considerable expertise in design and construction. The straight-line equivalent of the dome is the *pyramid*, and the same comments apply.

It is possible to construct a series of roofs side by side, so that they combine to form one large roof. All such arrangements are called *multiple spans*. When north lights are combined into a multiple span, they are sometimes called saw-tooth roofs.

Multiple-span roofs cannot be constructed to drain only to the edges. There will be runs of internal gutters, and these require special care in design, construction and maintenance. Any leak or overflow constitutes a major problem for the building's occupants.

Mention was made of *canopies* as examples of butterfly roofs, but any shape can be used as a canopy. A canopy is really only a roof without walls, and has a number of applications. Apart from petrol stations, the most obvious are loading bays at factories, and carports alongside houses. Grandstands are often no more than large canopies, usually with the back closed to increase the degree of shelter.

The list of names can never be complete, and there will always be a few roofs which cannot be described by any of the names given. This need not be a problem to us. We name standard items to make description easier, and if an innovative design is produced, it deserves a comprehensive description until such time as its use becomes widespread.

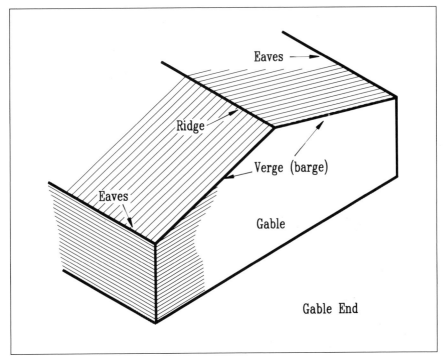

Figure 2.6 Terms associated with gables.

Just as we have names for standard roof shapes, we also have names for standard roof parts. Again, these are intended to simplify the task of describing a roof or feature, and they sometimes enable us to describe a roof shape which does not conform to any of the standard descriptions.

The first such feature is, strictly, not a roofing part at all! This is the *gable*. This can be visualized in the most simply shaped building, rectangular with a symmetric pitched roof, the most common shape for a house, and almost certainly the way a young child would draw one. The end walls rise to a point in the middle to follow the roof pitch, and these are the gables. The roof ends at the gables, and the gable detail is referred to as the *verge*, or sometimes the *barge*.

At the start of this chapter, and in Figure 2.1, mention was made of the eaves, which are normally the intersection of the roof slope and side wall. However, it is also possible to have *overhanging eaves*. This detail is more widely used in other parts of the world than in Britain, but is occasionally encountered here for some special applications.

Overhanging eaves offer certain advantages in extremes of climate. They can provide shelter from strong sunlight, preventing direct solar radiation from entering the windows. They can ensure that intense rainfall is thrown well clear of the building, and they give protection against falling snow by keeping a sheltered zone alongside the house. Of course, similar details can be applied at the gables to give *overhanging verges*, or *barges*.

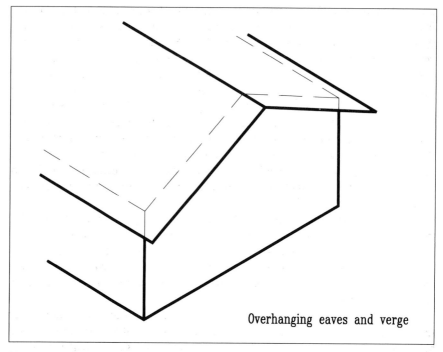

Overhanging eaves and verge

Figure 2.7   Roofs can overhang the walls of the building.

Overhanging eaves are an attractive detail for security purposes, widely employed in prisons. It is not easy to climb on to a roof which has overhanging eaves
unless a ladder is available. They are also useful in any application where there is a
risk of vandalism.

Parapets have long been a feature of roof design. Once they were simply used as
a safety barrier around the edges of high roofs, and are still used as such today around
the roofs of tower blocks, where the occupants have access to the roof as an extension
of their living space, or for recreational purposes.

However, about 20 years ago the use of parapets was extended for less practical
reasons. It became the fashion to design buildings as rectangular blocks, with flat
roofs (i.e. roofs with very small pitch). The planning authorities appeared to become
obsessed with these shapes, and demanded them for new factories and warehouses.
Unfortunately, the roofing materials which provided the very small pitches were
relatively expensive, and manufacturing industry could not afford extra costs in
building factories.

The compromise solution was the use of parapets. A roof could be constructed
from the usual materials (in many cases this was corrugated asbestos cement), with
roof slopes pitched at about 5°. As this was visually unacceptable to the planning
authorities, a parapet was constructed around the building to hide the pitched roof and
give the appearance of a flat top. In the case of factory buildings these parapets were
often 1m high or more.

The parapets were purely decorative; they made no contribution to the performance of the roof. Indeed, they could fairly be said to be detrimental to the roof function in that they compelled the use of internal gutters, with greater risks of water entering the building. It will be shown (Chapter 16) that parapets can also be responsible for snow accumulating into drifts; the weight of drifted snow is quite considerable, and this can become a design condition for the roof. In fact, it was the widespread use of parapets which led to structural problems caused by snow drifts. This encouraged research, which resulted in a better knowledge of roof loading due to snow.

The cost of roofing materials has changed, and other materials have become available. Parapets are expensive in that they require a supporting structure, and cladding to both front and rear. There are various modern roofing systems which can be laid to a very low pitch, sometimes as shallow as 1.5°. It seems more logical to spend money on a good-quality roof, than to economize on the roof materials and then incur costs in a parapet to hide the roof!

Another feature found in roofing is the *hip*. Hips are sloping corners of roofs, like the four corners of a pyramid. Their use can transform the appearance of a building.

One frequently used application for hips is to reduce the height of the gable wall. The triangular, upper section of the wall is then no longer required, and this means that there is no obvious difference between the sides and the ends, as both have a level top and similar gutter arrangement.

On tall buildings, low-pitched hipped roofs can be used to give the appearance of flat tops. This works well in city centres, where it is not usually possible to view the building from a distance. The roof becomes invisible, and, this is a better approach than using parapets.

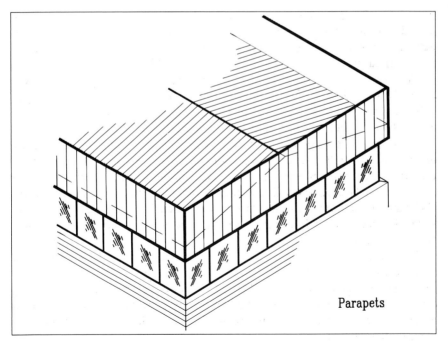

Parapets

Figure 2.8 Parapets can hide pitched roofs.

The economics of hips are not straightforward. There will be waste associated with the necessary skew cutting of roofing materials. Extra structural supports will be needed, following the line of the hip, and special weatherproofing measures may be required along the hip line. The gutter system must be continued across the end of the building, and there may be a need for extra rainwater pipes. To set against these additional costs, there is the saving in the peak of the gable wall. There will probably a further saving in the cost of the scaffolding, which does not need to be so high. The size of the costs and savings varies according to the building geometry, and the relative costs of the roof and wall materials.

Hips are also used whenever a building turns through 90° on plan, i.e. for L-, T- or E-shaped buildings. At each change of direction there is a hip on the outside of the bend. On the inside of the bend there is a *valley*.

The easiest way to visualize a valley is by comparison with a hip. A hip is sometimes said to be a sloping ridge, and this is a very reasonable description. Certainly a hip is at the top of a slope, and this is the chief requirement for a ridge. In the same way, a valley is at the bottom of a slope, at the bottom of two slopes in fact. So a valley is similar to sloping eaves, it is the place to which water will drain, and will require a gutter. Because the gutter slopes, it will drain very efficiently and very fast. Consequently, valley gutters are often much smaller in cross section than eaves gutters.

The term 'valley' is also used for the gutter between two spans in a multiple span condition. This kind of valley is level, so when there is a risk of confusion, the sloping variety is referred to as a *hip valley*.

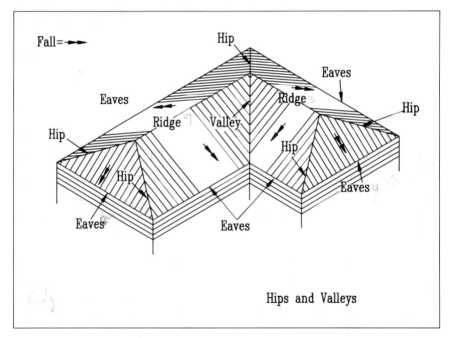

Figure 2.9   An illustration of hips and valleys.

We have mentioned the effect on the eaves of parapets, hips and valleys, but another detail can have a major influence on the appearance of the eaves line; this is the *curved eaves*. This detail is formed in profiled sheet and provides a curved transition from the wall to the roof slope. It is sometimes referred to as 'Floclad curves', but this is not strictly accurate, as Floclad is a trade name which specifically describes metal sheets, crimped into a curve. Curved eaves can also be created in smooth (non-crimped) curves in metal, fibre cement and GRP.

There is no best radius for curved eaves details, but the most popular radii are in the range 400–600mm. Very small radii are difficult to form, and large ones can lead to complications in the supporting structure. The original concept of the curved eaves was that rainwater would drain down the roof, around the eaves, down the wall, and into a drainage channel alongside the building. It was intended that the elimination of rainwater goods would provide savings to finance the extra cost of the curved sheets and ground works, and was believed that this would improve the long-term security of the building against leaks.

A number of buildings were built according to these principles, but certain problem areas were identified. The volume of water flowing down the walls during a storm can be very large, creating a waterfall effect over any windows or doors. Consequently doors or windows must be located in the gables, whether or not this suits the intended use of the building.

However, the aesthetic effect of curved eaves sheets led to widespread popularity, and architects deviated from the original concept in order to achieve curved eaves effects on inappropriate buildings. Today some of these deviations are so commonplace as to be generally accepted, with no thought as to the illogicality of the designs.

The most obvious of these is the factory shed type of building which must have numerous openings in its sides (loading doors, personnel doors and office windows). The occasional presence of waterfalls in front of these openings cannot be tolerated, and drainage channels are inconvenient additions across door openings.

The modern design solution is to introduce a gutter at the bottom of the roof slope, immediately up-slope from the curve. This gutter is not easily seen, and does not detract from the appearance, but introduces some unwelcome costs and risks. The gutter and rainwater pipes are a direct extra cost, and they create a need for extra structural supports. However, the risk of water ingress is far more important, as any fault in the gutter will now allow water to enter the building.

When curved sheets are used correctly, to create a continuous skin around the building, the roof and wall sheets must be identical because the profile is continuous. The sequence of construction is inflexible. The roof and walls must progress together at both sides of the building. This may increase the demands on labour, and on scaffolding costs.

There are many ways of fitting windows into roofs. However, true glazed opening windows are best fitted in very steep slopes, or vertical faces (as already mentioned under north lights and mansards). Various clever arrangements have been developed by which a vertical window can be accommodated in a sloping roof.

The best known of these is the *dormer*, which can be seen on any modern housing estate. The dormer is the means of making good use of the roof space. It increases

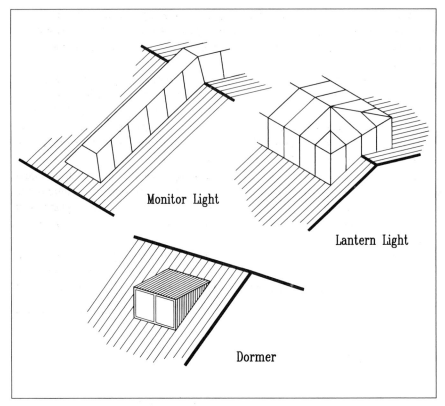

Figure 2.10 Some common forms of glazed rooflights.

the amount of headroom as well as providing light and ventilation. Dormers represent an easy way of increasing the amount of living space in a dwelling.

Factories and commercial buildings are unlikely to require dormers, but sometimes employ other means of locating windows in the roof. One such detail is the *lantern light*, which is positioned at the ridge. Essentially, the central 2–3m of roof is raised by about 1m. The top is glazed, and windows are also located in the vertical sides. Opening lights can be controlled by hanging cords, or by electric motors and switches.

Ridge lighting may not provide adequate illumination without the aid of side lighting, e.g. from windows in the side walls. An alternative arrangement, which can provide more uniform illumination, is the roof *monitor*. Monitors are raised strips of roof, in lines from ridge to eaves, and one or both sides of these strips provide the positions for the windows. The detail can be used as an alternative to north lights, in that the glazing can be arranged to face north so as to give uniform light without solar gain.

Monitors can have opening lights in just the same way as lantern lights. The monitors do not have to extend all the way to the eaves, in fact the aesthetic effect is better when they are curtailed a little way up-slope from the eaves.

Natural lighting can also be provided by way of glazing, or translucent sheets, in the slope of the roof, i.e. at the same pitch as the roof. In a domestic application this would be a *skylight*, a simple, small glazed opening to provide natural lighting to an attic or loft.

There are various other possibilities for large, single-storey buildings, such as factories or warehouses. These buildings depend on the rooflights for most of the lighting requirements. When the roof is constructed from profiled sheet, the most common detail consists of *profiled translucent sheets*, replacing a percentage of the normal profiled sheets. These have the advantage that the concentration of light can be varied to suit the intended use of the building.

Another possibility is the *barrel vault ridge light*. This is simply a translucent curved assembly for location at the ridge. Its width can be varied according to the amount of illumination required. It has the advantage that it does not disrupt the main roof slopes, and this can be a major benefit when the roofing is a site assembled, multi-layer system.

Of course, there are roof systems which are not based on profiled sheets, and which have a relatively smooth surface. These include felt, asphalt and single-membrane systems. Such roofs often employ *domed rooflights*, which consist of

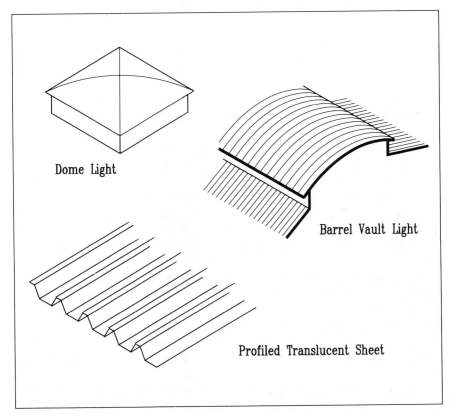

Figure 2.11   Some common forms of plastic rooflights.

either circular or square lights up to 2m across, fixed on a raised kerb. The design of rooflights, and their actual performance, is discussed in detail in Chapter 10. At present we are more concerned with identifying the different types.

Ventilators also occur in a wide variety of shapes, types and sizes. Many buildings enclose processes which produce excessive heat or unpleasant fumes, and so some form of ventilation is needed to maintain tolerable conditions within the building.

If the ventilation requirements are moderate it may be possible to provide sufficient air changes by means of opening lights in a north light, monitor or lantern light. Panels of louvres are sometimes used as an alternative to opening lights, and these may be manually operated or thermostatically controlled.

If greater ventilation is demanded, some form of purpose-made ventilator will be used. One popular arrangement is a continuous ridge ventilator; this is a form of box with flaps which may be open or closed. It can be 'natural draught' or 'powered'.

There are two advantages from the ridge location. Wind or air movement will create turbulent conditions around the ridge, and this will amplify any natural ventilation. It is a basic requirement of ventilators that they are located over a hole in the roof; it is easier to weatherproof this opening when it is at the ridge, because there is less water at the top of a slope.

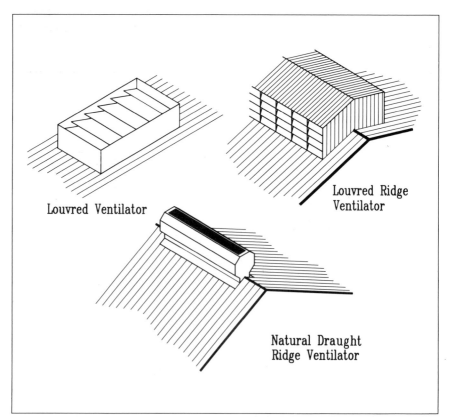

Louvred Ventilator

Louvred Ridge Ventilator

Natural Draught Ridge Ventilator

Figure 2.12   Some common forms of roof ventilators.

When the source of heat or fumes is clearly defined, it is possible to position the ventilator directly above the source. In such applications the ventilators are usually quite large, and are fitted with a powerful fan. If the opening is in mid-slope, it may be more difficult to weatherproof the edges. Ventilators are sometimes incorporated into the fire protection system, and this is discussed in more detail in Chapter 15.

People may need to walk on roofs for maintenance of roofing or ventilators, cleaning glazing, access to plant rooms, and annual inspections. Some roofing materials are brittle or easily dented, some are extremely strong, and the remainder lie somewhere between these two extremes. If there is a regular access route (e.g., to a plant room, or to glazing), and if there is any doubt about the ability of the roof to support such foot traffic without damage, then a permanent *walkway* should be provided.

Sometimes a roof walkway can provide a fire escape route when safety measures are demanded. Walkways are fitted immediately above the roofing, on some form of structure or brackets, so that the roofing does not have to support any load. Walkways are usually of lightweight construction, with a handrail on one side. They must be sufficiently durable to survive for many years in exposed conditions without undue deterioration.

Careful attention to detail is required where the supporting system penetrates the roof. There may be a large number of penetrations, and these must be weatherproof for the life of the roof.

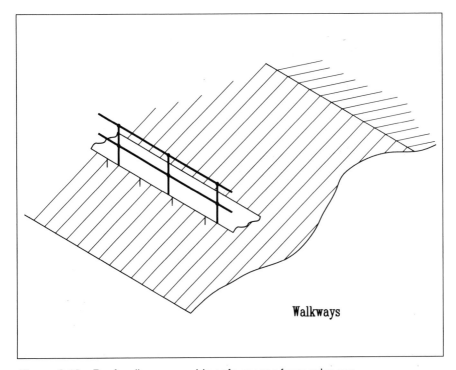

Walkways

Figure 2.13 Roof walkways provide safe access for regular use.

Large roofs may support enormous quantities of snow. British Standard BS 6399 suggests that this could be as much as a tonne on every 17m^2. If a mass of snow were to slide off such a roof and fall several metres, it would represent a very considerable hazard. This potential danger is sometimes addressed by the installation of *snow boards*.

Snow boards are simply barriers to restrain drifts of snow. They are fixed to the roof, close to the eaves. When drifted snow starts to melt, it usually melts from the bottom, because heat escaping through the roof thaws the frozen snow in contact with the roofing. This can break the bond between the drift and the roof, and lead to the whole drift sliding down the roof; the risk increases as the roof pitch increases.

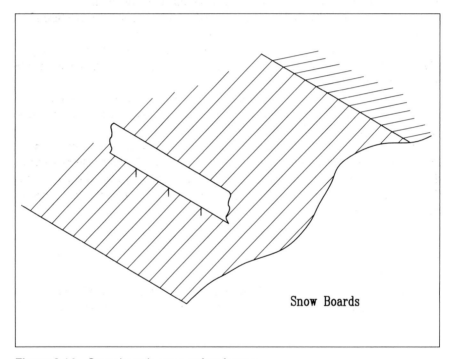

Snow Boards

Figure 2.14   Snow boards are a safety feature.

The snow board does not usually need to project very far above the roof, but a small gap between the boards and the roof must be left, through which the melting snow may drain.

Snow boards must be securely anchored if they are to restrain large masses of ice and snow. It is sometimes possible to fix them to the roofing material, but in most cases it is safer to fix them to the roof structure. The fixing brackets will have features in common with walkway supports, and will present the same requirements for weatherproofing.

The need for snow boards is greater in more northern countries, but should not be overlooked when designing roofs in Britain, especially when they are high and steeply pitched, and when there is public access close to the walls. Otherwise it will probably be sufficient to locate snow boards directly above doorways.

The snow boards are usually planks of treated timber, but can sometimes be formed in aluminium or steel.

Where roofs adjoin walls, or change direction, material, or slope, there is a requirement to protect the exposed edge, and render it weatherproof. In the great majority of cases this is done by means of a *flashing*.

The primary purpose of a flashing is to keep out the weather, but flashings also offer possibilities for aesthetic expression. There are several choices for material, many choices for colour, and infinite choices for shape. Of course, the flashing must always be designed to be weathertight, but if the appearance of the roof can be enhanced at the same time, then opportunity should not be ignored.

Flashings usually take their names from the associated details (ridge flashings, verge flashings, parapet cap flashings, etc.) — see Chapter 12.

Other features common to almost all roofs are guttering and rainwater pipes. These may be disguised or concealed, or displayed prominently, perhaps being made more conspicuous by virtue of shape or colour.

Rainwater goods are discussed fully in Chapter 17. The various types are described, and there are explanations of the methods for calculating the necessary sizes for given intensity of storms.

Just as the list of names of types of roofs can never be complete, neither can the list of roof features ever hope to include all possibilities. The descriptions given in this chapter include the popular items, which occur regularly in roof construction. Experience within the roofing industry will certainly reveal other less common examples to extend the roofing vocabulary.

STANDARDS
BS 6100:Subsection 1.3.2:1989 - Definition of terms for types of roofs,
  their related components and different types of roof coverings.

TILES, SLATES AND SHINGLES

3

Plate 3.1   Plain tiles are used to striking effect on this multi-featured roof.
(By courtesy of Redland Roof Tiles Ltd.)

Elsewhere in this book it will be shown that roof constructions may include elements providing thermal insulation, acoustic insulation, sound absorption, fire protection, decorative linings, vapour control layers, and various other features. However, the most important part of any roof construction is that item which keeps out the rain and snow. In Britain, rain may occur on any day of the year, and snow is possible for about half the year; wind speeds can be high, and the wind can be from any direction. The main purpose of a roof is to protect the contents or occupants of a building. (In other parts of the world with hot, dry climates, different priorities may of course prevail.)

There are several fundamentally different ways of providing the protective outer layer of the roof, sometimes called the *weather skin*; it is a measure of the importance of the weather skin that each generic form has its own chapter in this book.

This chapter is devoted to those systems which are applied to the roof in small units, and which provide weather protection by means of unsealed, overlapped ends and sides. Typically, the small units are *tiles*, *slates* or *shingles*.

These systems, whilst being individual and different from each other, nevertheless have certain features in common. Because the roof is formed from small overlapping units, it can only be made weatherproof by virtue of a pitch which provides rapid run-off of rain water. Typical pitches are in the range 17.5° to 45°, but steeper pitches are possible. These pitches are steep by comparison with those required by profiled sheets, membranes and fully supported standing seam metal, all of which are discussed in later chapters.

Tiles, slates and other natural stone products, although heavy by comparison with most other forms of roofing $(30–80kg/m^2)$ are often the most aesthetically pleasing of all roofing products. A wide variety of colour and texture is possible, there are numerous shapes and sizes — particularly for tiles — and the pattern of overlaps can create other visual effects. Aesthetics are important for this roof form because the steep pitches make it likely that the roof slopes will be visible.

It follows that tiles and slates are a popular roofing choice for houses. Most home owners prefer to have a reasonably steeply pitched roof as this provides extra storage space or, sometimes, extra accommodation within the roof void. Houses have short-span roof trusses or rafters, usually in the range 6–10m so the weight of the roof covering can be supported by simple roof trusses. Tiles and slates are not usually a viable roof covering for large clear-span factories; their weight would necessitate massively strong trusses, and these would be uneconomic. However, it is possible to use mansard construction with tiles or slates on the steep slopes, and a lightweight, less attractive finish on the upper slopes.

Chapter 16 explains how roofs are loaded by snow and wind. It is shown that wind gusts can exert considerable upward forces, attempting to tear away the roof covering. There is also an explanation as to how the wind force varies with the roof pitch; the forces are greatest at the lowest pitches. It will be appreciated that the use of heavy roof covering, at steeper pitches, reduces the quantity of mechanical fixings required, and simplifies the fastening requirements.

Chapter 14 describes acoustic performance of roofs. Weight is shown to be the most important property for insulating against sound transmission. Tiles and slates provide this weight, and the loft void, created by the steep pitch, enhances this sound insulation effect.

The aesthetic benefits of tile and slate roofs are important to home owners. People take a pride in the appearance of their own home, and their own district. They are far less concerned about the general appearance of their place of work. However, the most important property of tiles and slates in respect of domestic roofing is their durability. A new house is usually purchased on a 25-year mortgage; borrowers would not accept a roof which required replacement before the loan was repaid. Tiles and slates are extremely durable, in fact one tile manufacturer offers a 100-year guarantee on his product.

By contrast, many factories and commercial buildings are built for a relatively short life; the scheme only proceeds if the investors are confident of an early return on their money. Sometimes the roof is designed to last 15 or 20 years, it is accepted that replacement may then become necessary, but the replacement will be financed out of profits already earned by the building.

Table 3.1   A comparison of tile and slate products.

Product	Minimum Pitch°	Typical Weight Kg/m²	Typical Size
Plain clay tile	40	70	265 x 165
Plain concrete tile	35	80	265 x 165
Interlocking clay tile	20*	40	420 x 300
Interlocking concrete tile	17.5*	45	420 x 330
Large Welsh slate	20	35	600 x 300
Small Welsh slate	35	45	300 x 150
Natural stone	45	80	Various
Bituminous shingles	15	10	Various
Pressed steel	10	7	Various

* Consult manufacturer's technical literature.

Table 3.1 provides a brief summary of the various products. The figures are approximate only, and are intended to be used as comparisons between the different systems. There are many manufacturers of roof tiles, many slate quarries, many forms of natural stone, differing qualities of workmanship, and different levels of exposure. It is therefore impossible to provide a specific answer as to minimum roof pitch for a product type, or even its weight per square metre. Manufacturers will provide specific recommendations for their own products.

Before giving detailed descriptions of any particular form of tile or slate, it is necessary to make a few comments about the method of support of tiled or slated roofs. The individual units are laid on *timber battens* to which they may be fixed by means of nails or hooks, or by purpose-made clips. It is not always necessary to fix every unit, sometimes it is enough to fix alternate tiles or some other such arrangement.

The battens may be considered to act as structural members spanning between the rafters. Their size is influenced by their span, the loading, the properties of the timber, and the requirements of the fasteners. Fortunately, the sizing of battens can be carried out by reference to BS 5534: Part 1. While a range of batten sizes is possible, 38 x 25mm and 38 x 19mm are the most popular for normal applications.

The battens are located within the roof at a position which may be inaccessible for inspection or maintenance. It is desirable that they should match the weather skin for durability. It is therefore sensible to treat the battens with an effective, long-term preservative. Some care must be exercised in the selection of the preservative in order to ensure the intended result. Some preservatives can be aggressive to some metals, and this could lead to the fasteners being attacked by the preservative.

For example, some excellent timber preservatives are based on salts of copper. It would be unwise to use aluminium nails in timber which had been so treated: the nails could become corroded, and this could lead to roof failure. Copper nails would be perfectly suitable, however, and there are other preservatives (e.g. creosote) which are not injurious to aluminium fixings. In fact, it is unusual for fixings to be corroded because of the timber preservative, as water from roof leaks or condensation would also need to be present. However, the risk should be acknowledged and avoided.

The battens are used in conjunction with *underlays*. An underlay fulfils two main purposes: it provides a second line of defence against the weather, and it helps to resist wind loads by forming a diaphragm between the internal and external air pressures.

It is quite possible that a weather skin, consisting of hundreds of small overlapping units, may have a few poor laps, or damaged tiles. At these positions there may be small gaps through which rain or snow could be driven. Similarly, such gaps may occur in a roof which was initially perfect; this could be the result of structural settlement, or of accidental damage from foot traffic or from a severe storm. In any of these events, the presence of a continuous and intact underlay prevents water from entering the house.

The way in which the underlay can help to resist wind loads can be explained by reference to an example. Imagine a shed with a large door; there is a severe gale and the door is facing the prevailing wind. As the wind swirls around the shed it creates suction, or uplift, forces which attempt to tear off the roof. Now suppose the door is opened, wind rushes into the shed causing a build-up of internal pressure; this force is also attempting to lift the roof. The internal pressure, in conjunction with the external suction may be critical, and it is possible that the combined effect of the two forces may be enough to cause structural damage, even though neither force was sufficiently strong on its own.

In just the same way, the total force on a tiled roof is the result of external suction combined with internal pressure. In fact, it does not require a large door to cause significant internal pressures; buildings are not airtight and various gaps can contribute to the effect. A suitably installed underlay can resist the internal pressure forces, thus leaving the tiles and their fixings to cope with the external suction.

The underlay, which is sometimes referred to as *sarking felt*, is laid so as to sag slightly between the rafters. This creates a clear channel down which water can drain, to be discharged at the eaves (Figure 3.1).

In Scotland it is standard practice to fix tongue and groove boards (or sometimes plywood or chipboard) over the rafters, these are also called sarking (or sometimes

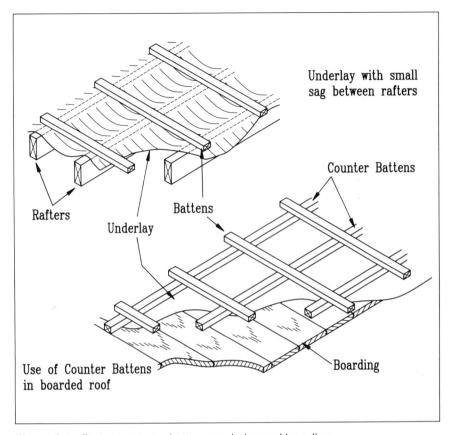

Figure 3.1   Battens, counter battens, underlay and boarding.

'rigid sarking'). The underlay is laid directly on top of the boards. Battens are required to maintain a space between the underlay and the tiles, but the battens may not be laid directly on the underlay as this would create traps to catch moisture, and that would promote rot and decay.

The usual answer is to use counter battens running from ridge to eaves; these are located over the rafter positions. The counter battens lift the battens clear of the underlay and create clear drainage runs. There is no ideal size for counter battens, but it is usual to use smaller-sized timber than the tile battens. Boarding and counter battens are not exclusive to Scotland, being used elsewhere for higher-quality roofing and in very exposed conditions. It is also possible to place the underlay over the counter battens so that they receive greater protection. Battens and counter battens on a boarded roof are shown in Figure 3.1.

The traditional material for underlays was hessian-reinforced bitumen felt. The reinforcement was necessary to prevent tearing. Recently other materials have emerged; these include heavy-gauge reinforced PVC and polyethylene. To some extent the choice is a matter of user preference, but there are some guidelines as to desirable qualities.

An underlay must be durable. It is intended to last the lifetime of the roof, and cannot be replaced without removing the roof. It must be waterproof in that it resists saturation (but should allow the passage of water vapour), and must not be weakened or rotted by water. It must have adequate strength against tearing around the nails. It should not make loud noises when moved by varying pressures (i.e. it should not sound like a flapping sail).

At the eaves, a small portion of the underlay emerges, and is lapped into the gutter. The underlay material must resist ultraviolet radiation, or this exposed part will be degraded and much of its value will be lost. The material should also be resistant to attack by birds or rodents, and should be compatible with any timber preservative used on the rafters or battens.

Some underlays incorporate a metal foil. This can reduce heat lost by radiation, and can create an extremely effective barrier against water or vapour, but it can also weaken radio or television signal strength. This becomes very important when an internal aerial is intended.

The above comments refer to the underlay after installation, but certain other properties are useful during the actual fixing sequence. The material must be capable of being handled and fixed in normal weather conditions (roofs are exposed, and there is almost always some wind or breeze). The size of rolls must strike a balance between a good rate of cover and a comfortable weight to handle. The surface should be capable of accepting some form of marking, as an aid to setting out for the tiles or slates. The edges and surface of the material should not be so rough as to injure the user's hands; however, they should not be so smooth as to become slippery in wet weather.

There have been several recent developments in this field. One new idea is to use galvanized steel trays, from ridge to eaves, as a sarking. The battens are fixed to upstand edges, and the trays may be filled with insulation. It is claimed that this reduces the number of rafters, is quick to fix, and makes for a robust construction.

Another development is the use of profiled sheets, in compressed bitumen and fibre, as a combined underlay and boarding. The battens are fixed directly over the crowns of the profile. This arrangement can be used to provide shelter very rapidly during construction, it is a second line of defence against storm damage, and it can enable tiles and slates to be used at a lower pitch than would otherwise be possible. It can also contribute to thermal and acoustic insulation.

Whatever system is chosen, the outer covering of tiles is the main defence against the weather. There are two possible types of tile, *plain* (double-lapped) or *interlocking* (single-lapped), and each type is available in clay or concrete.

Clay is the traditional material for tiles, but concrete tiles have been used for about a 150 years. Concrete tiles are less expensive and consequently have a far greater share of the market.

Both materials offer some advantages. Clay is available in attractive natural colours, the colour is permanent, and is continuous through the whole of the tile. Concrete tiles can be given a coloured textured surface, but this can fade or be lost due to weathering or abrasion. A wide range of through colours is also available to produce more permanent effects, and these are to be preferred.

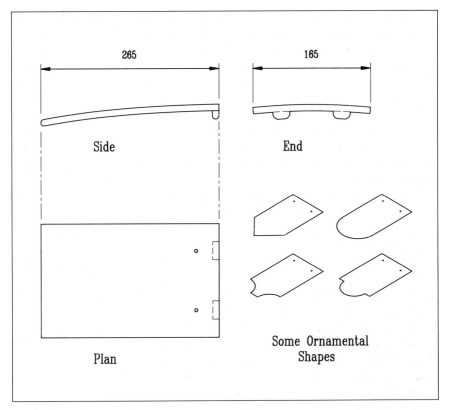

265

165

Side

End

Plan

Some Ornamental
Shapes

Figure 3.2   Plain roofing tiles.

Concrete can be produced with great uniformity under controlled factory conditions. Clay is dependent upon the clay and the firing process, and may have more variation of properties. In particular, some clay tiles can be more porous than concrete, are also more vulnerable to frost damage, and may be more irregular in shape. Consequently clay tiles must often be laid to a steeper pitch to ensure rapid and efficient run-off of rain water.

Concrete is a little heavier than clay, because it is densely packed in manufacture. However, the difference in weight is not large, and is a very small proportion of the design load.

A typical plain tile is shown in Figure 3.2. The standard size is 265mm long x 165mm wide. The tile may have a slight curvature, or camber, in both length and width, to reduce the possibility of capillary action between the tiles — the lap may be a close fit at the edges, but the tiles are further apart within the lap.

It is usual for the tiles to have two nail holes, and one or two nibs at the upslope end, or head. A nib may be continuous or intermittent, and a typical arrangement is illustrated in Figure 3.2.

Plain tiles are also produced with shaped tail ends to create decorative effects. Figure 3.2 includes several well known examples. Some manufacturers offer an impressive range of ornamental shapes. Tiles with semi-circular ends create a 'fish scale' effect; other shapes produce different aesthetic results. Special patterns can be further enhanced by using more than one colour of tile, but these arrangements are for spectacular decoration, not for weather protection.

The tiles are fixed to the roof by nails, or by purpose-made clips. Whichever sort of fastener is used, it is vitally important that the correct fastener material is chosen. The tiles may last for 100 years; the fasteners should be equally durable, otherwise the life of the roof may be drastically shortened by the premature failure of its least expensive component.

The traditional fasteners for tiles were galvanized steel nails, which have now been superseded by many materials which are more durable (aluminium, copper and stainless steel are all good choices), but it is essential to check that the timber preservative in the battens is not aggressive to the tile fastener material. The length and gauge of the nails must be properly matched to the tile type and specified batten.

BS 5534 provides guidance on the frequency of fixing. It is necessary to fix the first two rows of tiles at the eaves and ridge, at valleys, hips and abutments, and the first few tiles from the verges. In the remaining areas some rows may be laid without fixing depending on the exposure of the site, so the specification could call for 'fixing at every third row', or 'fixing at every fifth row'. In exposed areas it may be necessary to fix every tile.

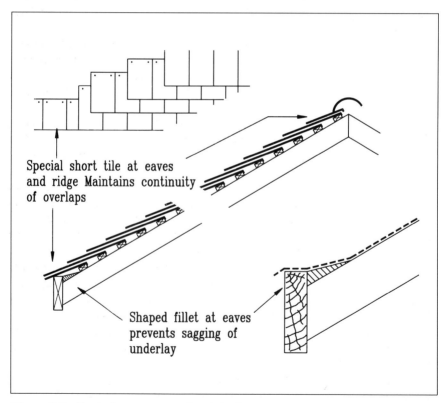

Figure 3.3 Typical arrangement for plain tiles.

The overlapping arrangement is shown in Figure 3.3. The joints are staggered in successive rows so that the side of a tile in one row will be directly above the centre of a tile in the next row above or below (broken bonding). Special tiles are produced, one and a half times the normal width; these are used to start alternate rows so that the stagger is automatic.

The endlaps are substantial; the top edge of a tile is covered, not just by the tiles in the next row upslope, but by the bottom edges of the tiles in the slope above that. The overlap of a tile to another tile in the next row but one is called the *head lap*, and must be at least 65mm. For exposed conditions, a greater head lap is needed, perhaps as much as 90mm. Special, short tiles are produced to maintain the overlaps at the eaves and ridge.

Figure 3.3 shows how tiles are lapped on a roof. It will be noted that the tiles do not lie at the same angle as the rafters: the downslope end of the tile is raised up by two thicknesses of tile. It is the actual pitch of the tile which determines the performance of the roof, but roof pitches are always described in terms of the pitch of the supporting structure or rafters. This is logical, as different tile thicknesses and cambers would affect the pitch of the tiles.

The shaped timber fillet at the eaves is very important, as it ensures that the underlay cannot sag behind the fascia board. Any water running down the underlay must be drained to the gutter; ponds formed in sagging underlay would inevitably leak into the building at a lap or nail hole. Many otherwise sound, tiled roofs have failed because this detail was overlooked.

At a verge, or gable, the tiles are usually seated in mortar. This is an efficient way of closing the irregular gap between tiles and brickwork; it also helps to ensure that the tiles are firmly held down, and not vulnerable to wind damage.

Because plain tiles are small, they can be made to follow curved roof structures. This is not true of the much larger interlocking or single-lap tiles.

As with plain tiles, interlocking tiles are available in both clay and concrete. Again concrete is heavier than clay, but the weight of these tiles, as laid, is only about half that of plain tiles. This is purely because there is much less overlapping material.

The interlocking edges of the tiles are designed to resist penetration by rain, so the tiles may be used on lower pitches. The minimum slope shown in Table 3.1 can only be an indication; there are countless designs for interlocking tiles, and the performance of each varies with exposure. In particular the headlap must be at least 75mm, and this may have to be increased at low pitches. It is always necessary to study the manufacturer's recommendations when establishing a specification.

Figure 3.4 shows a typical interlocking tile. The edges create a labyrinth effect and some interlocking clay tiles have a corner notched to prevent an unsightly accumulation of thicknesses. Figure 3.4 also shows some popular forms of interlocking tiles. Often the choice of a particular shape is simply a matter of personal preference.

Interlocking tiles are supported on battens, in much the same way as plain tiles. However, the battens are far more widely spaced, and this may mean that they are more heavily loaded and need to be of a greater size.

Figure 3.5 illustrates the typical lapping arrangement. The system shown has the sidelaps in line (straight bond), but some systems use staggered side laps in the same way as plain tiles (broken bond).

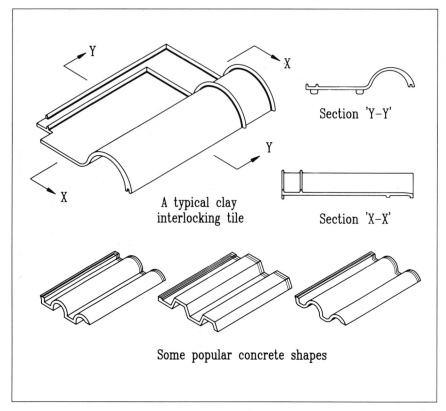

Section 'Y-Y'

A typical clay
interlocking tile

Section 'X-X'

Some popular concrete shapes

Figure 3.4   Profiled roofing tiles.

The arrangement for the fixing of interlocking tiles is usually more critical than for plain tiles, and there are several reasons for this. Interlocking tiles are large, so they attract a greater wind load. They are used on lower pitches, and this further increases the wind loads (the effect of slope on wind load is described in Chapter 16). Because of the reduced lapping, there is less dead weight to hold the tiles down.

Not only are the uplifts greater, but the shape of the tile can reduce the possibilities for nail holes. In any case, nail holes are not welcome in the headlap labyrinth which is intended to prevent water ingress. Furthermore, the size of the tile means that there can be a considerable leverage effect as a tile attempts to rotate about a single-top fixing or double-top fixings.

The ideal solution is to fix the tile, close to its tail (downslope edge), without introducing a penetration in the form of a nail hole. This solution is possible by virtue of purpose-made clips. Typical forms of these clips are shown in Figure 3.6. There is a standard clip for the main area of the roof, and special clips for use at the eaves or verges where local wind effects can be particularly severe.

BS 5534 offers advice on fixings. In general, interlocking tiles should be clipped when the pitch is under 30°, and at least nailed for pitches over 45°. At pitches between

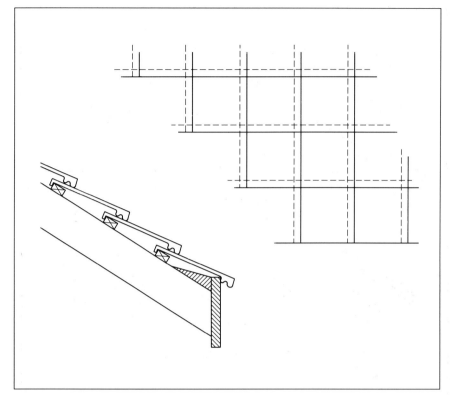

Figure 3.5   Typical arrangement for profiled tiles.

30° and 45°, either method or both may be used. For pitches over 55° it is usual to both nail and clip. It is not always necessary to fix every tile. Depending on the loading, pitch, exposure, tile shape and tile size, it may be possible to omit some fixings, but it is always necessary to fix every tile at the eaves, ridge, hips, valleys, abutments and verges. The remaining tiles are divided into two zones, local and general; either may be fully nailed, fully clipped or clipped in a chequer pattern, depending on the calculations from BS 5534.

Modern housing developments consist of one- and two-storey, steep-roofed dwellings in wide streets. The buildings are set back from the street by the width of the front garden, so the roofs are extremely visible, and form an important part of the visual impact of the houses or streets. The sympathetic selection of different tile profiles, colours and textures can make a major contribution to the overall appearance of an estate.

Slate is a natural material, and varies according to the locality in which it is quarried. Welsh slate is rightly famous for consistent quality, and for its availability in large unit sizes. The stone splits cleanly, and can be produced as thin, lightweight units.

The best-quality slates are impervious to water and strongly resistant to attack by

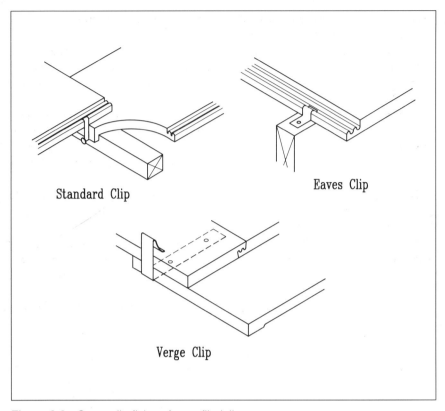

Standard Clip

Eaves Clip

Verge Clip

Figure 3.6 Some clip fixings for profiled tiles.

atmospheric pollution. Their colour does not fade. They offer the possibility of extreme durability. For example slates on the roof of St Asaph's Cathedral, were still sound after 250 years.

Welsh slate is often supplied in a range of thicknesses; 5.0mm, 6.5mm and 8mm being the usual choices. The thicker gauges offer greater strength for those applications which are likely to be heavily loaded.

Some other forms of slate are used in greater thicknesses, but smaller units; they may have a more attractive appearance. Imported slates should only be used when full test data is available; it is possible for slate to perform well in Mediterranean countries, but be unable to withstand the severe frost which can occur in Britain.

When slate is used in small units, it is nailed at the top and fixed in a manner very similar to plain tiles. Of course, slates do not have nibs, so every slate must be fixed to prevent it sliding down the roof.

Large slates may be fixed in a slightly different way. Instead of being nailed at the head, they are nailed at the centre. This reduces the leverage effect, as there are lengths of slate on either side of the fixing. Figure 3.7 shows the typical arrangement of slates for centre nailing. The top of each slate ends at the centre of a batten, and the fixing for the next slate is immediately upslope from the top edge of the previous slate. A course of short slates is used at the eaves to ensure that all joints are protected, and these short slates are head nailed.

The arrangement requires that the batten is wide enough to provide a bearing for one slate while leaving room for the fixing for the next slate. This is not feasible with narrow battens, and it is recommended that the minimum batten width should be 38mm.

The slates are delivered to site without nail holes. The holes are formed on site by the slater, using the spike on the back of a slater's axe. This is a skill which comes from practice. The slate is turned upside down for spiking, some material breaks away from the upper surface as the spike penetrates, and this forms a countersunk detail for the nail head.

Nails must be in a durable material: aluminium, copper and silicon-bronze are the preferred choices. A nail-free system based on purpose-made stainless steel hooks is also possible for sheltered areas.

Extra-wide slates are available to start alternate rows, and thus set up the staggered joint arrangement. At the verges the slates are bedded in mortar, as are plain tiles, and it is good practice to allow the slates to project up to 50mm beyond the wall line — this gives additional protection against leaks.

Slates are such simple shapes that it is inevitable that they should be offered in

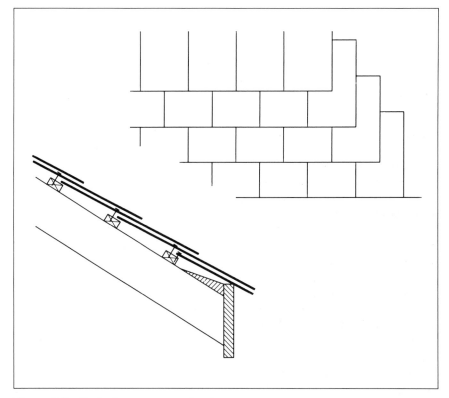

Figure 3.7   Typical arrangement for slates.

other, synthetic materials. Fibre cement slates are lighter than real slate, and less durable, but can provide an inexpensive alternative where extreme durability is not of paramount importance. Some synthetic slates are difficult to distinguish from the natural material from the normal viewing distance.

Synthetic slates should always be fixed in accordance with the manufacturer's instructions, but these will probably be similar to those for real slate.

Some interlocking tiles are virtually flat, and are coloured blue-grey. Although applied as tiles, they can give an overall effect which is reasonably similar to slates.

Other forms of natural stone are sometimes used for roofing. Some forms of sandstone and limestone are capable of being split into reasonably uniform units for roofing. These split stones are used in greater thickness than slate, and in relatively small units. They are heavy and only suitable for steep pitches. As the stones are often irregular in thickness, considerable skill may be needed to incorporate them into an effective roof.

Bituminous shingle is another roofing material which is fixed in small pieces, and which has a similar appearance to slates and plain tiles. There are various proprietary products, and Figure 3.8 shows a typical example. The composition is usually fibre-reinforced bitumen, with sanded surfaces. The sanded surface provides protection

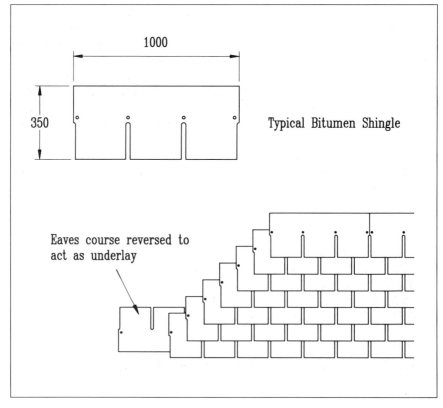

Figure 3.8 Typical arrangement for bitumen shingles.

against ultraviolet radiation, and can be coloured in order to offer a range of choices.

Shingles are laid on a continuous deck of plywood or chipboard. The fixings are usually galvanized or copper nails, but other arrangements are possible. Some versions incorporate an adhesive strip in addition to the nails. There are also torch-on varieties — this expression will be discussed in more detail in Chapter 6. Essentially it consists of a lower layer which is melted by a torch (e.g., a butane burner) and immediately placed in position. The molten adhesive solidifies to give a continuous bond.

Shingles are light in weight, and can be fitted to pitches as shallow as 12°. The durability of shingles will certainly not match slates or tiles. A reasonable life for a shingle roof would be about 20 years, but there are many applications where 20 years is sufficient, and where savings in initial cost are attractive. There are also applications where lightness is of prime importance.

At low pitches it may be necessary to use a continuous bitumen felt underlay as extra weather protection. All details should be designed by reference to the manufacturer's recommendations. It is wrong to assume that all products are the same: there may be a component or constituent which significantly changes the performance of one product relative to another.

There are also a number of products which consist of coated steel or aluminium, and which have been pressed into panels with the appearance of profiled tiles. These are of light weight, and are capable of spanning between purlins or widely spaced battens. Their useful life is likely to be around 20 years for steel, or 30 years or more for aluminium.

Profiled metal tiles are mainly used for refurbishment projects, particularly where the existing structure is suspect and a reduction in dead weight is desirable. Manufacturers will supply details of their products, and advise on details, fixings, etc. Profiled metal systems are fully described in Chapter 4. They are mentioned here because they are produced and marketed as an alternative to traditional tiles.

Cedar shingles imported from Canada, are used for aesthetic effect or where low weight is important. They are fixed in a similar manner to head-nailed slates. When suitably treated they can provide a service life of up to about 25 years.

STANDARDS
BS 402:1990 - Clay roofing tiles and fittings.
BS 473:1990 - Concrete roofing tiles and fittings.
BS 5534:Parts 1&2 - Code of practice for slating and tiling. (Design)
BS 8000:Part 6:1990 - Code of practice for slating and tiling. (Craft)

FURTHER READING
The David & Charles Manual of Roofing - John H. Wickesham - 1987

PROFILED SHEETS

4

Plate 4.1   Concealed-fix aluminium roofing incorporating hips and hidden gutters.
(By courtesy of Alcan Building Products.)

Profiled sheets are made from a wide variety of materials. Although a flat sheet of
thin material has very little structural strength, forming the sheet into a series of ribs,
with crowns (high points) and troughs (low points), can greatly increase the strength.
This strength enables the sheets to span between purlins without the need for
continuous support. The most common profiles are corrugated and trapezoidal, but
there are also many special shapes. The special shapes are usually concealed fix
systems which are unique to individual manufacturers.

The sheets are delivered to site as relatively large units. Typical widths are around
1m, but the lengths vary according to the material, and this will be discussed later.

The sheets are fixed on the roof with the profiles running from ridge to eaves, so
that the troughs perform as continuous drainage channels. Joints are formed as *side
laps* and *end laps*. Side laps are formed at the high point of the profile, so that
rainwater is shed to either side. End laps have the uppermost sheet overlapping the
downslope sheet by a relatively small amount, and this requires that the roof is
pitched in order that the water drains properly, without entering the laps.

If the sheets were to be overlapped with dry joints, in most parts of Britain it would
be necessary to use roof pitches of 20° or more to keep out the weather. However,

it is possible to install the sheets with compressible seals in the overlaps; this enables pitches to be reduced to about 5° for most systems.

Lightweight profiled sheets, at low pitch, are particularly suitable for wide-span industrial buildings. In fact, profiled sheets are as widely used in industrial applications as are slates and tiles in domestic applications.

The sheets are attached to the purlins by fasteners; these are usually screws with washers, but can also be nails or rivets in some applications. Corrugated sheets must be fixed through the crowns, as the curved shapes would be too difficult to seal in the troughs. Trapezoidal sheets are made up of flat surfaces, and so sealed fixings are possible in both troughs and crowns, but some specifiers insist on crown fixing. Concealed-fix systems are designed to be fixed without penetration of the weather skin.

In theory it is possible to design a structure with purlins at any positions. In practice the purlins are usually spaced at around 1.8m. This has much to do with structural stability of the rafters and trusses, and is not greatly concerned with the properties of the profiled sheets. It is likely that the majority of buildings will continue to have purlins at 1.5 to 1.8m centres.

The most popular materials for the manufacture of profiled sheets are listed in alphabetical order and discussed.

Aluminium alloys are durable and light, and have reasonable strength. Their performance can be adjusted by the introduction of very small quantities of alloying elements, and their strength can be modified by tempering. In most applications,

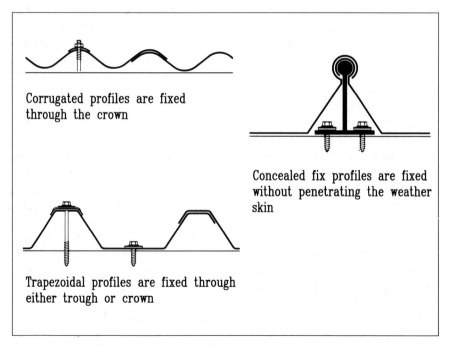

Figure 4.1 Positions of fixings in profiled sheets.

aluminium alloys may be used as plain, mill-finish metal, but coloured finishes can be applied for decoration.

The aluminium sheets may be roll-formed to any length which can be handled with convenience. (Essentially, roll-forming consists of passing flat metal through shaped rollers to produce a profiled sheet.) Alternatively they may be pressed to shape, and their length will be governed by the length of the press.

Mill-finish material is initially bright silver metallic, but the surface dulls with time as a layer of oxide forms. This oxide protects the metal from corrosive influences, and is extremely stable. Weathered aluminium roof sheets are usually a light metallic grey, the exact shade varying according to local atmospheric pollution. If the surface is scratched, bright metal is exposed, but the exposed metal will dull as a new oxide layer is formed. The sheets may be supplied with a smooth finish, or a textured finish which is described as 'stucco embossed'.

There are three ways in which colour finishes may be applied to aluminium. By far the most common is by means of organic coatings; there are numerous coating materials including PVF_2, polyurethane, polyester, alkyd amino and polyester powder. It is also possible to anodize the metal; this consists of first thickening the oxide layer, then colouring the oxide with permanent dyes. Anodizing gives a superb, durable, metallic finish, but this is expensive, and is not often necessary in roofing applications. Finally, there is conversion coating, which is really an etch priming treatment which is sometimes used as a low-cost method of dulling the new metal (e.g. near airports where dazzle must be prevented).

The greatest advantages of aluminium are its durability, its light weight and its environmental benefits. Aluminium roof sheets can be expected to last 40 years or more, with little or no maintenance; this is a typical lifespan for an industrial building. The metal is little affected by marine conditions, and is particularly suitable for use in a coastal environment. Its light weight places less demand on hoisting equipment, and is unlikely to be critical when sheets are stacked on the roof prior to fixing. Most aluminium is smelted by means of hydroelectric power; this is a renewable energy source which does not pollute the environment. Aluminium is an ideal material for re-cycling, as the energy to recycle is much smaller than the energy to extract new metal from its ore.

Two less desirable properties of the metal are its rate of thermal expansion, and the possibility of bimetallic corrosion. All metals expand on heating, and contract on cooling, but aluminium expands and contracts more than most. Design details for aluminium roofs must take account of thermal movement; the usual allowance is 1mm of movement for every 1m of length. Bimetallic corrosion occurs when two dissimilar metals are in contact in damp conditions — an obvious example is aluminium sheets on steel purlins; corrosion is easily prevented, however, by plastic insulating tapes between the metals, or by a thick coating of suitable paint. Contact with copper or lead must be avoided; even rainwater which has washed copper can be harmful to aluminium.

Fibre Cement sheets were the most popular choice of industrial roofing material for a large part of the twentieth century. The sheets were of corrugated form, and consisted of cement reinforced with asbestos fibre. The product declined in popularity when some forms of asbestos were identified as being injurious to health. Synthetic fibres have been developed to replace the asbestos, but the first attempts were not completely successful, and this further damaged the reputation of the product.

There is a widespread belief that the current fibre cement products have overcome the earlier difficulties, and perform as well as the asbestos-based products. However, they are still very new, and confidence will only be fully restored when the sheets have proved themselves in the field over a reasonable period.

The profiles are factory-produced by specialist equipment. To make a new profile requires substantial investment, so the choice of shapes is very limited. Sheet lengths are not great; 3m is usually the maximum. In fact, the material is quite heavy, and 3m is about the limit which a man can handle with comfort; a steel sheet of the same weight would be twice as long, and an aluminium sheet five times as long.

Fibre cement is a durable material, and can remain virtually unchanged for many years. It can be adversely affected by some industrial pollutants, and by marine

Table 4.1 A comparison of profiled roofing products.

Material	Profile Depth[1] mm	Material Thickness mm	Weight Kg/m^2	Safe Temperature °C	Life[3] years
Aluminium (plain or coated)	35–40	0.9	3	-80 to +100	40+
Fibre cement[2]	80–100	6.0	15	-25 to +200	30+
GRP	30–40	1.0–1.5	2	-20 to +100	15+
Steel (galvanized)	30–35	0.7	7	-20 to +200	5+
Steel (galvanized and coated)	30–35	0.7	7	-50 to +80	15+
Steel stainless	30–35	0.7	7	-200 to +650	40+

Notes
(1) Depth and thickness are typical for sheets spanning 1.8 m between purlins, and supporting average British loadings.
(2) The properties for fibre cement are based on lengthy experience with asbestos cement; they may not be valid for other types of fibre reinforcement.
(3) The life quoted is the the typical period which may be expected if maintenance is neglected. The life of sheets can usually be extended by cleaning, repainting, repair, etc.

atmospheres. In such cases the surface is softened, and the exposed fibres become dirt traps.

The material is slightly porous and can absorb some water from rain or condensation. This moisture, in combination with dirt, can promote the growth of moss and algae. When the surface has been softened by local pollutants, the effect becomes more pronounced.

It has been observed that the material becomes more brittle with age, and less able to support impact loading. Fibre cement sheets do not deform under load, like metal; structural failure can occur with very little warning. It is therefore essential to use crawl boards, or roof ladders, when working on fibre cement roofs. Walking on new sheets is foolhardy, and walking on old sheets is almost suicidal.

New fibre cement sheets are pale grey in colour, with a dull matt surface. In time the sheets become stained or discoloured by the local environment, or change appearance because of the growth of moss or algae. The sheets are of constant composition throughout their thickness, so scrapes and scratches have little detrimental effect.

Fibre cement can be factory coloured, or site painted; this may be required for decorative reasons, or to create harmony with the close surroundings. Such colours are seldom bright, and usually have a very low level of gloss. They can protect the surface against deterioration in aggressive conditions. The colour finishes deteriorate more rapidly than colour finishes on metals, and site painting may become necessary after about ten years if the colour finish is to be preserved.

Fibre cement has lost much of its market share to metals; this is because it does not have any significant advantages in today's roofing industry. However, it is durable and easy to use, is not affected by the same pollutants which attack some metals, and is less likely to cause 'drumming' during rain or hail storms. Its absorbent surface can be an advantage in some applications where condensation forms on the back of the sheet.

Its weight can be a disadvantage in handling, and in stacking on roof structures. It is brittle and easily damaged, and cannot support foot traffic. It is not popular on aesthetic grounds because there is limited choice of profile shapes, and of colours.

Glass Reinforced Plastic (GRP) is mainly used for translucent roof lights (Chapter 10). However, the material is also used for some specialist applications such as soakers and curved eaves sheets, and for that reason is also included in this chapter.

The sheets consist of polyester resins, reinforced with glass fibres. They are not as strong as metal sheets of similar profile, but their thickness can be increased to provide some compensation. It is usual to allow greater deflections in GRP sheets than in those of other materials. The fixing screws require washers of greater diameter than would be used with, say, metal. The sheets are not capable of supporting foot traffic. GRP has a relatively high rate of thermal expansion, similar to that of aluminium, and details require special care when using long sheets.

Various surface finishes are possible, but to achieve the level of durability described in Table 4.1, Tedlar PVF film is required. Periodic cleaning of the sheets extends their life, and this can be further increased by site painting, with a suitable coating material, when the surface has become roughened by natural weathering.

Galvanized mild steel is the base material for numerous coated products, but can also be used uncoated for certain short- and medium-term applications.

Steel is available in a vast number of alloys and tempers, but the most common material for profiled roof sheets is Z1 material. This is a little stronger than the most popular aluminium alloys, and three times as stiff. In effect this means that if matching profiled sheets were formed in steel and aluminium of equal thickness, the steel sheet would support a little more load, before collapse, than the aluminium, but the aluminium sheet would deflect, under load, three times as much as the steel.

Mild steel could not be exposed to the weather in an unprotected form; it would rust very rapidly from the action of rain and condensation in conjunction with atmospheric oxygen. The usual protection is galvanizing, which consists of dipping the steel in molten zinc to produce a thin, continuous coating of zinc on the surface of the steel. For roofing applications, the most frequently used specification is G275; this means that there are 275g of zinc per square metre of steel (this is the total, so there are 137.5g on each of the two surfaces). This is equivalent to saying that the thickness of the zinc coating is about 20 microns ($20\mu m$).

Zinc is said to give a self-sacrificial protection to mild steel. If the zinc is scratched, so that the steel becomes exposed, it would be natural to assume that the protection was lost. However, the zinc around the scratch would migrate to cover the exposed steel, thus reinstating the protection. The chemistry of this reaction is the same as that which produces bimetallic corrosion in aluminium. Of course, the zinc coating over scratches will never be as thick as the original damaged coat, but even a very thin layer of zinc will provide protection for as long as it lasts. The durability of galvanized steel roof sheets is therefore much the same as that of zinc; once the zinc protection has gone, the steel cannot survive for long.

The durability of zinc in Britain is well documented thanks to the Ministry for Agriculture, Fisheries and Food (MAFF), which set up a test programme to investigate the rate of corrosion of zinc throughout the country. A grid of 10 km squares was established, a sample located in every square, and samples regularly taken in and checked.

The result of this investigation is published in the form of the MAFF map of United Kingdom Atmospheric Corrosivity Values. The map uses a colour code to show the rate of attack in each square of the grid. The lightest corrosion rate quoted is about $13g/m^2$, and the heaviest $50g/m^2$. At the heaviest rate, 137.5g of zinc would be consumed in a little less than three years; some improvement can be achieved by thickening the zinc coating, but there is a limit to this, as roll-forming will crack the zinc if it is too thick.

The maximum rate of attack of zinc coatings occurs in a small area around Leeds and Bradford, presumably because of the pollution of industrial Lancashire falling with Pennine rain. Other areas with high rates of corrosivity are London, the Midlands, Lancashire and Tyne and Wear. The map also highlights a narrow coastal strip where corrosivity is higher due to the marine influence.

Galvanized steel is a robust material, but has too short a useful life for many applications. Where greater durability is required, some form of protection must be specified.

Organic coated galvanized mild steel This is a general-purpose roofing material which enjoys a major share of the industrial roofing market. There are many possible organic coatings, including PVF_2 and polyester, but by far the most popular for roofing is plastisol. This is a plastic of the PVC family, it is applied in thicknesses of up to 200 microns (200μm), and usually has a leather-grain finish.

Plastisol protects the zinc from atmospheric attack, and is sufficiently thick to provide good protection against abrasion. Other organic coatings are much thinner, and do not provide such good protection; however, they may have better decorative qualities, and could be preferred in less aggressive environments.

Plastisol can provide 15 years' protection to galvanized steel roof sheets, and this can be increased by sensible selection of colours and details. After 20 years or so, repainting may become necessary to extend the life of the sheets. The colour of the plastisol may change with time, but this will usually be a reasonably uniform change over a complete roof slope.

Roll-formed sheets are cut to length with a special guillotine or shear; this exposes a bare metal edge. The exposed steel is protected by migration of the zinc, but this is likely to be the first area where rust appears. It is recommended that the cut edges are coated with a protective paint when the environment is aggressive.

Plastisol is not tolerant of high temperatures, it can be damaged by temperatures of 80°C, or higher. Dark-coloured roofs can attain temperatures of 80°C, whereas pale colours are unlikely to rise to more than 60°C. It is therefore necessary to restrict plastisol-coated roof sheets to pale and medium colours.

For reasons of economy, it is normal practice to limit plastisol coatings to the weather side of steel sheets. The reverse side is usually given a thin coating of organisol or polyester, which is adequate for normal conditions. However, care must be exercised when this backing coat is exposed, for example at an overhanging eaves. In such cases it is good practice to site paint the exposed coating, or to order steel sheets with plastisol to both front side and reverse.

Steel sheets are, typically, about half the weight of fibre cement and about double the weight of aluminium. Their rate of expansion and contraction is about half that of aluminium. Their durability does not match that of aluminium, but is considered to be sufficient for many applications.

Minority Products

Stainless steel is a form of steel with a high chromium content. It can be used unprotected, and possesses exceptional durability. This is not a common material, and it is difficult to cut and drill. It is mentioned here in order to present a complete picture, but expert advice should be sought in producing a specification or design.

Another minority product is *Bitumen impregnated corrugated sheets*. These are formed from a matrix of reinforcing fibres which are impregnated with bitumen at high temperature and pressure. They are usually formed into relatively shallow profiles which are suitable for close-spaced purlins. They have limited durability, and are mainly used for agricultural buildings and temporary construction.

There can be little argument that the preferred colour for profiled industrial roofing is light grey. This is the colour of weathered mill-finish aluminium and of plain galvanized steel; it is also the colour of natural fibre cement. Furthermore, it is by far the most popular colour of plastisol on galvanized mild steel.

Pale grey is certainly a suitable colour for such applications. Not only do pale colours help to preserve plastisol, but they help to conserve energy by radiating away less of the building's heat in winter; similarly, they promote greater comfort in summer by absorbing less solar radiation. It is interesting to note that mill-finish aluminium is much more effective than pale grey paint in this respect, even when the aluminium is old and weathered.

Whatever the material, profiled sheets are fitted to the roof with overlapping edges to keep out the weather. Figure 4.2 shows some variations on the side lap theme.

Most side laps include a sealant, usually a preformed mastic strip which adheres to the sheets, and is sufficiently compressible to close uneven gaps. The sealant is located at the highest point of the lap so that, should rain be driven into the lap, it will flow out again. Some sheets have an underlap edge detail which provides a drainage channel in the event of failure of the seal.

Side laps usually require screws or rivets to hold the lap closed. The holes for these are drilled on site, with the sheets in their final position on the roof. This operation carries a risk that the underlap edge may be deformed, and this would reduce the effectiveness of the lap. In an effort to counter this, many sheets are offered with 'supported side laps' (Figure 4.2).

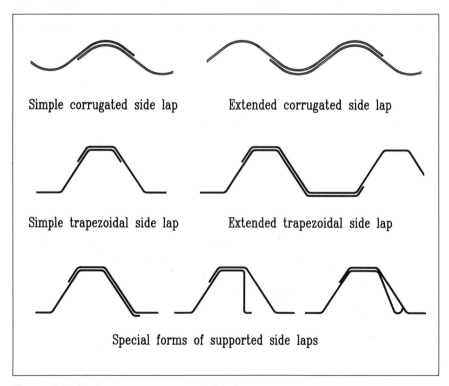

Simple corrugated side lap   Extended corrugated side lap

Simple trapezoidal side lap   Extended trapezoidal side lap

Special forms of supported side laps

Figure 4.2 Various arrangements of side laps.

Whenever possible, profiled sheets are fitted in single pieces which extend all the way from the eaves to the ridge. However, this is not likely to be possible on really long slopes as the sheets would become too heavy, difficult to handle, or prone to excessive thermal movements. In all such cases, the solution is the introduction of endlaps. The sheet profiles are identical, and the material is relatively thin, so one sheet fits very closely inside the next. Sealants are used to give greater weather protection.

Figure 4.3 shows two possible end lap arrangements. The simple version is usually used for steel sheets, but the expansion lap is preferred for aluminium. An expansion endlap requires a wide bearing at the purlin, and it is usually necessary to increase the width of the purlin flange by means of a *ledger rail*. Although expansion joints are more common with aluminium, it should not be overlooked that steel also expands and may sometimes require end laps which allow movement.

The number of seals, and their location, offers considerable scope for discussion and argument. There is no doubt that a single continuous seal will provide total weather protection for as long as it remains intact; modern cross-linked butyl mastics are extremely durable, and should remain intact for 20 years or more. Unfortunately, the standard of site workmanship is not always high, and many specifiers believe that a second seal provides additional protection.

Some manufacturers of steel sheets state that the laps should exclude water, whether rain or condensation; they do not want the thin backing coat to be exposed

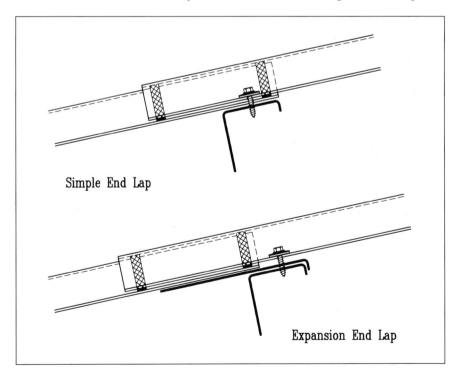

Figure 4.3 Two types of end laps.

to aggressive conditions. Their preferred solution is to employ narrow overlaps (e.g. 75mm), and to use a wide sealer strip which fills the whole width of the lap. Other steel-sheet manufacturers prefer to use two separate sealer strips.

Manufacturers of aluminium sheets recommend wider endlaps to accommodate thermal movements; the width of overlap is often as much as 200mm. Wide single seals would not be practicable for these overlaps and, in any case, aluminium profiles are more flexible than steel profiles; it is doubtful whether such wide seals could be compressed properly.

When two seals are used, there is agreement that one should be close to the weather end of the lap (i.e. the downslope end), as it is obviously sensible to exclude water from the overlap joint. The position of the second seal is often at the centre of discussion.

According to one school of thought, the second seal must be downslope from the fixing screws; otherwise, it is argued, a failure of the first seal can result in roof leaks before the second seal has a chance to operate. Certainly it would be possible for water to enter the building by way of the fixing holes. However, some manufacturers of steel sheets argue that the second seal should be close to the upslope end of the overlap. They reason that there may occasionally be condensation on the back of the sheets, which could run down and fill the lap, and attack the sheets through the thin backing coat.

Manufacturers of aluminium sheets are less worried about the possibility of corrosion due to condensate, and more concerned about compressing the seals. They prefer to place the second seal directly under the fixing screws, where the clamping action is greatest. This is not possible in expansion joint end laps, however, as the screws do not penetrate both sheets.

Figure 4.3 shows two end laps; in each case the second seal is shown at the upslope end. This in no way suggests that this is the 'best' arrangement, it is simply one possibility. There are reasons to commend each of the three possibilities, and a user should always seek the advice of the manufacturer of the particular sheet which he proposes to use.

At the beginning of this chapter concealed-fix systems were mentioned, and one example was illustrated in Figure 4.1. These sheets are formed in aluminium or steel; they have complex shapes which allow the fixings to be hidden inside the ribs, and which makes them difficult to end lap. They are usually fixed in continuous lengths from ridge to eaves, which enables them to be fixed on very shallow pitches, perhaps as low as 1°. This would create a problem with thermal movement, but for the many ingenious methods which have been devised for fixing these products. These methods include sliding clips, rotating clips, flexible purlins and slotted holes.

Each concealed-fix system is a unique product; a prospective user should read the appropriate product literature, and seek the advice of the manufacturer. In some cases Agrément Certificates provide independent assessment.

For about 20 years there has been a fashion for curved eaves sheets to produce a modern, or streamlined, effect. Curved sheets have been produced in most materials.

Fibre cement and GRP sheets can be made in smooth curves, to virtually any radius. However, there is a significant cost in tooling up for a given radius and roof pitch, and as a result the products are offered in a few standard sizes only. It is always possible to commission a special product for very large projects which justify the development costs.

Steel and aluminium sheets can be curved by forming crimps across the profile. This is a very versatile system as the tooling is simply that required to form a crimp in the particular profile, but the tool can produce infinite variations. Each crimp bends the sheet through approximately 4°, but this can be controlled between 3° and 5°, by varying the pressure during bending. The total angle of bend is governed by the number of crimps, and the radius is a function of the crimp spacing.

It is possible to produce curved sheets down to about 300mm radius; if these are eaves sheets they will have about 20 crimps. These figures will vary a little according to the profile depth and shape.

Curves of very large radius, say over 750mm, require special care; the crimps become widely spaced, and the curved appearance can be lost as the sheets assume the visual aspect of a polygon.

Profiled sheets are available in such a wide range of materials and shapes that this chapter can offer no more than a brief introduction to the subject. Fortunately, manufacturers are only too pleased to promote their particular products; they will

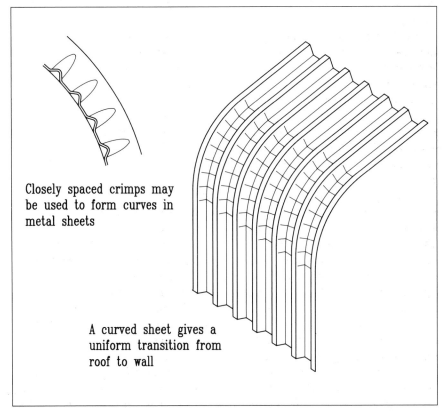

Closely spaced crimps may be used to form curves in metal sheets

A curved sheet gives a uniform transition from roof to wall

Figure 4.4   Curved profiled eaves sheets.

almost always provide comprehensive literature covering their own profiles. This product literature will be invaluable to anyone wishing to learn more about the design and application of these products.

STANDARDS
BS 5427:1976 - Code of practice for performance and loading criteria for profiled sheeting in building.

FURTHER READING
Profiled Sheet Metal Roofing and Cladding: A guide to good practice - NFRC Publications - 1991

One Voice
One Aim
One Standard

The MCRMA represents the major UK manufacturers in the metal roofing and cladding industry and seeks to foster and develop a better understanding amongst specifiers and end users alike of the most effective use of metal building products, components and systems.

As part of its commitment to a policy of technical excellence the MCRMA has initiated and funded a number of research programmes on industry-related topics. In addition, the association has produced a series of technical design guides which has proved invaluable for a wide range of specifiers throughout the building industry.

The MCRMA works closely with its European counterparts to ensure that there is European wide agreement on standards, a co-ordinated response to new developments and to ensure that the views of UK metal sheet producers are heard.

THE METAL CLADDING & ROOFING MANUFACTURERS ASSOCIATION LTD.
18 Mere Farm Road Noctorum Birkenhead Merseyside L43 9TT
Telephone: 051-652 3846 Fax: 051-653 4080

FULLY SUPPORTED METAL

5

Plate 5.1   The magnificent copper dome of the London Planetarium.
(By courtesy of The Tussauds Group Ltd.)

The use of metal as a continuous weatherproof membrane is a long established practice. The particular advantages of this form of roofing are that metals are totally impervious to water, they have the strength to resist damage (e.g. from foot traffic), they can be extremely durable and they can also confer aesthetic benefits.

The metal is selected to provide the most suitable outer skin for the roof. Structural support is provided by a continuous deck, often tongue and groove boards or chipboard. The deck spans between rafters or purlins, and carries all loadings from snow, wind, and foot traffic. These loads are transferred directly through the metal, hence the description 'fully supported'. This chapter will concentrate on the metal top skin; the deck is outside the scope of this book.

Although continuously supported metal roofs have a long history of use in Britain, they have never taken a major share of the popular roofing market. They have, however, always been seen as a high-quality product offering durability and security as a return for an initially high investment. In some other countries, this form of roofing is much more widespread; the visitor to Stockholm, for example, is surprised to find that almost all the roofing is in continuously supported metal. As materials and ideas spread throughout the EC and EFTA, it may be expected that expertise in continuously supported metal roofing will be imported into the UK.

There are, essentially, six metals which may be used for a continuously supported metal roof. They are aluminium, copper, galvanized steel, lead, stainless steel and zinc. Each has particular strengths and weaknesses, and none of the six can be claimed to be 'best'. Of course, it is quite possible that one will be the most suitable choice for a particular application. This chapter will attempt to make useful comparisons between these metals, and to show how each is attached to the roof. It is hoped that this will enable the potential user to make an appropriate selection.

Table 5.1 lists the six metals, and summarizes some of their more important properties. The typical gauge for roofing applications is shown, and lead is seen to be used in much greater thickness than the other metals. From the gauge and the density, the weight of the roofing may be calculated; lead is heavy and aluminium is very light, but the remaining metals have similar weights.

All six metals have significant rates of thermal expansion and contraction. For plain metal, it is reasonable to assume that the extreme service temperature will differ from the temperature at fixing by up to 50°C. For lead this would produce a change of length of about 1.5mm in every metre. For galvanized steel this would be little more than 0.5mm, and for the other metals about 1mm. Design details must take account of these thermal movements. (Note that coated galvanized steel, in all but the palest colours, is likely to absorb more solar radiation and thus become a few degrees hotter than plain metal. Hence its lower coefficient may be partially offset by a greater temperature range.)

The melting points are given to show the wide variations between the different metals, but it is unlikely that this property would ever be critical in a roof design. Even

Table 5.1   A comparison of properties of popular roofing metals.

	Aluminium	Copper	Galvanized Steel	Lead	Stainless Steel	Zinc
Typical gauge, mm	0.8	0.6	0.7	2.5	0.5	0.8
Density, tonnes/m^3	2.7	8.9	7.8	11.4	7.9	7.2
Coefficient of thermal expansion x10^6	23	17	12	29	17	22
Melting point,°C	660	1080	1350	330	1400	420
Thermal conductivity W/m °C	170	355	60	351	51	30

the melting point of lead is far higher than any likely operating temperature for a roof.

Thermal conductivity varies dramatically between the six metals, but is not of great importance, as separate insulation will always be required if significant thermal retention is demanded. The conductivity value is sometimes needed when carrying out detailed analysis of heat flow through the elements of a roof, or assessing the severity of cold bridges.

The table does not include any direct indication of strength; this would be difficult as the metals are available as a variety of alloys, and each may be offered in a range of tempers. The typical gauge takes account of the strength of the alloys and tempers in popular use.

There is no indication of price. This is because price is not fixed (like density or melting point). Prices of metals rise and fall relative to the prices of other metals or other commodities. The weight of the roofing material can be calculated from the typical gauge and the density. The current trading prices of aluminium, copper, lead and zinc can be obtained by reference to quoted prices on the London Metal Market — these figures are published daily in the leading newspapers.

The table cannot include durability, as there are too many factors and variables to consider. However, it may be assumed that aluminium, copper, lead and stainless steel are all extremely durable; they are all capable of providing a life in excess of 100 years. Any attempt to make general comparisons between them, in terms of durability, would be meaningless. Zinc is known to deteriorate more rapidly than those metals already mentioned, but can certainly provide a life of over 50 years in a rural atmosphere (though this can be significantly reduced in polluted industrial atmospheres). Galvanized steel is only suitable for short-term applications, as its life is likely to be in the range 5-10 years, according to local conditions. An organic coating on galvanized steel can prolong its life to as much as 30 years, provided details follow recommended methods.

In terms of this particular form of roofing, metals may be categorised as 'soft' or 'hard'. Soft metals are easily made to follow the supporting deck, or formed into seams and welts. Hard metals have a springiness, and are more difficult to form. Lead is a soft metal, as is high-purity aluminium. The remaining four metals are hard. Certain alloys of aluminium, in hard temper, may also be used as hard metals, but they are more frequently used as profiled sheets, spanning between purlins.

Metals are also classified as ferrous and non-ferrous. Ferrous metals are based on iron, for example, mild steel and stainless steel. Non-ferrous metals are often more expensive, but do not require protective coatings (as does mild steel); they usually have a high scrap value, and are readily recycled.

Metals can be corrosive to other metals; this is important when selecting fasteners, or fitting lightning conductors. The corrosion is initiated when two dissimilar metals are connected by a conductor of electricity. The conductor is usually water with dissolved impurities, and this is available on most roofs in large quantities! It follows that incompatible metals will always be at risk in roofing applications.

Fortunately, not all metals are incompatible. In fact metals may be placed in groups, the metals in each group being reasonably compatible:-

(i)   magnesium and high-purity aluminium
(ii)   cadmium, zinc and aluminium alloys
(iii)   iron, lead and tin
(iv)   copper, chromium, nickel and stainless steel.

Metals in different groups are likely to be corrosive towards each other. For example, aluminium and zinc can be used together without difficulty; aluminium is also suitable for use with galvanized steel as long as the zinc coating remains intact, but if the mild steel becomes exposed it will set up corrosion of the aluminium. Copper is very aggressive to aluminium alloys, and to pure aluminium, so the two metals must not be used together. Even rainwater flowing from copper to aluminium can cause rapid corrosion - but flow from aluminium to copper is not a problem.

It might be expected that stainless steel would be corrosive to other metals, but this is not found to occur in practice. Stainless steel forms an oxide which is a good electrical insulator, and this inhibits the corrosive action, so stainless steel is often the preferred fastener material with other metals.

This type of corrosion occurs because a galvanic cell is formed, and it is therefore termed *galvanic corrosion*. Because two metals are involved, it also known as *bi-metallic corrosion*.

The six metals will now be discussed in turn.

(i) *Lead* is the most popular metal for fully supported roofing applications in Britain. It has a long and distinguished history of service, and there is a reservoir of skill in its forming and fitting. It is perhaps most commonly known as the preferred roofing material for churches and cathedrals.

Table 5.1 shows the typical gauge as 2.5mm, although gauges from 1.8mm to 3.55mm are used, the exact gauge being determined by the conditions of use, exposure, etc. Because lead is subject to large thermal movements, it is advisable to limit the length of individual pieces; the usual recommendation is 1.5m maximum for 1.8mm gauge and 3.0m for 3.55mm gauge. The joints must include provision for movement.

Lead roofing is quite heavy; in the gauges mentioned its weight will be between 20 and 40kg/m^2. This weight is almost enough to counter the wind suction loads on many roofs, and may explain why lead roofing was used very successfully long before the magnitudes of wind suction loads were known or understood.

The traditional method of fixing lead to a roof is by means of *wood-cored rolls*. The arrangement is shown in Figure 5.1. The wood-cored rolls are in the line of the pitch, i.e. from ridge to eaves. They are spaced from 500 to 750mm apart, depending on metal gauge and loading conditions. The shaped timber cores are securely fixed to the roof deck.

The lead sheet is dressed around the core from one side, and fixed to it by means of copper nails at the highest point, and in the upper one-third of the panel only, so as not to restrict thermal movement. It has been found that copper, brass and stainless steel are compatible with lead, and make the most durable fasteners.

The second sheet of lead is dressed around the roll to cover the first piece and its fasteners. A flat strip at least 40mm wide is formed at the overlap side. It might be supposed that water would be drawn into such a detail by capillary action; this is

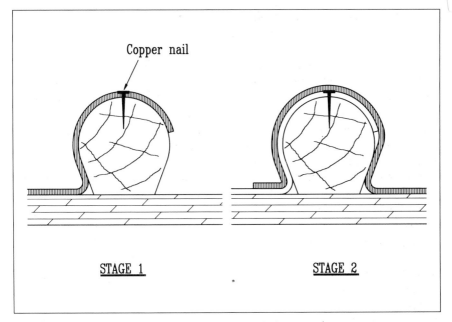

Copper nail

STAGE 1   STAGE 2

Figure 5.1   Forming a wood-cored roll in lead roofing.

indeed possible, but is not of any great consequence. Capillary action can draw water into but not through, a joint (but see the comments on thermal pumping and wind pumping later in this chapter.)

Apart from the fasteners, and the self weight of the metal, the lead is held down by a 'dovetail' action. The metal cannot easily be pulled off the cores because of their shape — the neck is much narrower than the widest part of the core.

Expansion or contraction between the rolls is not very important. The distance is small and the total movement is less than 1mm, and this is accommodated by bending at the rolls. Thermal movement in the length is a more important consideration. The sheets are up to 3m long so the movement could be almost 5mm. The joints must not restrict this movement or there will be large thermal stresses which could buckle or tear the metal.

On steep roofs (say 20° pitch, or more) it is possible to use simple overlaps. The guiding principle is that the length of overlap should give a vertical rise of at least 75mm. If the pitch is 20°, the length of the overlap is 220mm; if the pitch is any lower, the lap becomes unacceptably long.

Figure 5.2 shows a typical overlap joint for a steep roof. The free end of the sheet is retained by means of copper clips. These prevent the end from being lifted by the wind, and this ensures that a weathertight lap is maintained, without restricting movement. The top end of each sheet is secured by copper nails, and these are covered by the next overlapping piece.

For shallower pitches, overlaps are not an effective method of jointing. The preferred method is to use drips; these turn the roof into a series of steps or cascades.

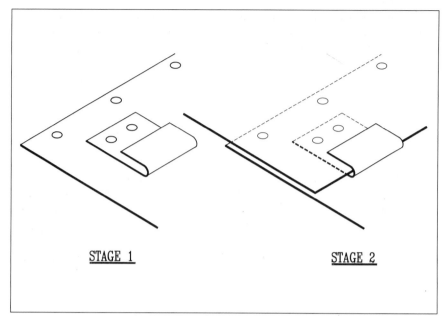

STAGE 1   STAGE 2

Figure 5.2   Forming a simple overlap in a steeply pitched lead roof.

The rolls must be terminated, and restarted at each step.

Figure 5.3 illustrates some of the ways in which the drips may be formed. The simplest details require the highest steps, while it is possible to make the steps as low as 40mm if some form of anti-capillary detail is incorporated. Again, the underlapping sheet is secured with copper nails.

Metals are not laid directly on to the supporting deck; some form of underlay is recommended. The underlay isolates the metal from any timber preservative which may have been used on the deck; it also provides a form of slip plane to allow unimpeded thermal movement.

In the case of lead, there is another important function for the underlay, however. Lead is dissolved, slowly, by distilled water, and condensation is distilled water. So condensation on the back of the sheet can have a corrosive effect on the metal, and could eventually lead to perforation. The traditional underlay was inodorous felt, an impregnated flax felt; this allowed the back of the sheet to 'breathe', and reduced the risk of local wet spots. However, as insulation levels have increased, the risk of condensation has become greater - condensate is present more often and in greater quantities. The recommended modern underlay is needled, non-woven polyester textile, which is resistant to rotting and will not adhere to the metal or the deck; it will allow some passage of air, and this can assist the drying out of any condensation.

Underlays also make a small contribution to thermal insulation, and a more significant contribution to acoustics, in that they help to deaden any drumming sounds from rain or hail.

Joints in lead may be soldered or welded (this is usually termed 'leadburning');

temperatures for such processes are relatively low, and do not constitute a major site hazard. These methods should not be used for joining large sheets, or thermal movements will become difficult to control. However, they are excellent for sealing around openings, e.g. at vent pipes, or skylights.

As lead ages, it develops a silver-grey patina; this patina is stable, uniform and insoluble, and is generally considered to be an attractive feature of the metal. Unfortunately, the patina does not form uniformly on new metal, and has an undesirable transition stage. Water on new lead sheets will produce an uneven deposit of white lead carbonate; this is most unattractive, and can be washed off by the rain to stain walls or other materials.

The solution to this is the use of patination oil. This oil is applied uniformly to the new metal, where it improves the initial appearance, prevents the formation of the white carbonate, and encourages uniform weathering of the lead surface. As roofs receive the most critical scrutiny when they are new, patination oil is strongly recommended.

(ii) *Aluminium* is the only other 'soft' metal in common use for fully supported roofing. For this purpose it is used in the form of a high purity alloy (at least 99.5% pure), in fully soft temper.

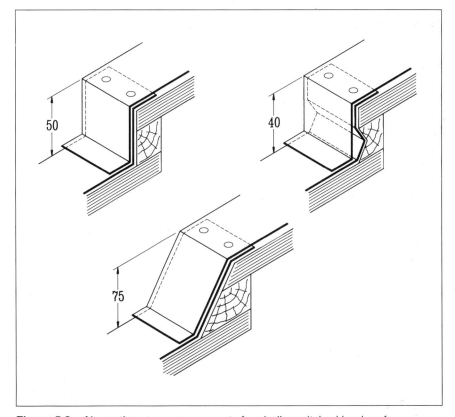

Figure 5.3 Alternative step arrangements for shallow pitched lead roofs.

Table 5.1 highlights a few of the differences between aluminium and lead, the most striking being the density. A typical lead roof sheet is around 2.5mm thick and weighs about 30kg/m², whereas an aluminium roof sheet is 0.8mm thick and weighs about 2kg/m². As the aluminium roof has less dead weight to hold it down, the method of fixing must be stronger, and is not possible to use wood-cored rolls as for lead. An alternative type of batten roll is often chosen. This is again formed around shaped timber, but the anchorage of the metal is more positive.

Figure 5.4 shows how a batten roll is formed, and how it is attached to the roof. This construction does not restrict thermal movement. The aluminium sheets are usually supplied in widths of 450mm, and this governs the spacing of the batten rolls. The preferred spacing for the batten rolls is 390mm.

As with lead, thermal movement must be accommodated; it is not possible to use long sheets, or to rigidly join short sheets. For very shallow pitches, say 5° or less, it is necessary to use drips, similar to those for lead, but using aluminium nails instead of copper! Aluminium has a slightly lower coefficient of expansion than lead, so the drips may be a little further apart; 4m is a practical maximum.

At steeper pitches it is possible to use welted joints as shown in Figure 5.5; the double welt is used for pitches up to 40°, and the single welt for steeper pitches. The welts must be formed in such a way as to allow a few millimetres of free movement for expansion. Welts can only be used because the metal is of a thin gauge; the six

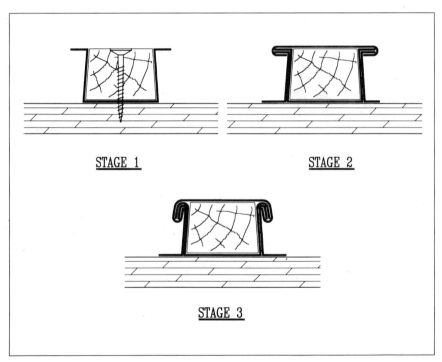

STAGE 1   STAGE 2

STAGE 3

Figure 5.4 Forming a batten roll in aluminium roofing.

plies in a double welt would amount to only 4.8mm, but six plies of lead would build up to an unacceptable 15mm.

At pitches of 20° or less, it is recommended that the edges of the sheets are coated with boiled linseed oil prior to forming the welts. It will be seen, from Figure 5.5, that fixing clips are incorporated in the welts. The usual arrangement is to place a single clip midway between the batten rolls.

Welts are less tolerant of thermal movement than are drips. Consequently the maximum spacing for welts is usually taken to be 3m. Aluminium is easy to weld, and so welding is the usual means of dealing with special details such as intersects and penetrations.

Aluminium sheets must not be placed directly on the deck (timber decks may have been treated with copper-based preservative which would be corrosive to aluminium). They are normally laid on an underlay of inodorous felt, or of needled, non-woven polyester.

Aluminium does not develop as dark a patina as lead; nevertheless it eventually weathers to a metallic grey, which is generally considered to be attractive.

Aluminium is work-hardened during fixing. This means that the act of forming batten rolls, or welts, increases the temper of the metal. If a mistake is made, the metal cannot simply be flattened out and reworked, as its temper will have been increased

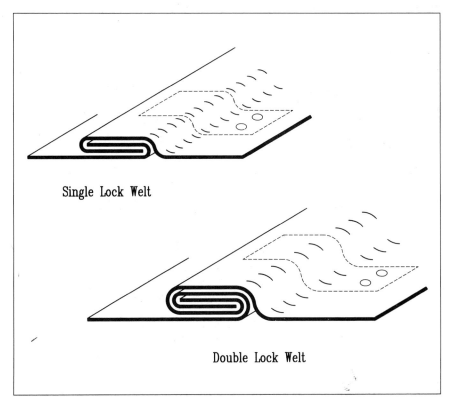

Single Lock Welt

Double Lock Welt

Figure 5.5 Single and double welts.

during working, and again during flattening. It is possible to soften the metal by heating it to a predetermined temperature, and allowing it to cool, but this requires skill and experience. Lead is virtually unchanged during working, and can be flattened and reworked without difficulty.

Aluminium can be an economical alternative to lead, and has been preferred in applications where security surveillance is difficult. Lead was often stolen from church roofs for its scrap value, but aluminium came to be preferred because 2kg of aluminium did not have the same value as 30kg of lead.

(iii) *Copper* is the most commonly used of the so-called 'hard' metals. It is associated with quality and durability — Sir Christopher Wren wanted to specify copper as the roof covering for the dome of St. Paul's Cathedral, but was overruled by Parliament on grounds of cost.

Copper has a relatively low coefficient of expansion, and this allows it to be used in slightly longer panels than is possible with lead or aluminium. It may be used in very thin gauges, and this influences the design of details. Its greater strength and hardness make it a little less easy to work on site.

The metal may be applied with batten rolls and drips, as described for aluminium. It may also be fixed in greater lengths, up to about 10m, by means of standing seams on expansion clips (Figure 5.6). The two-piece clip allows for rigid connection to the structure, with free movement of the seam. The seams are usually spaced about 500–600mm apart.

Of course, this same arrangement could be used with aluminium (in rather shorter lengths), or with lead; however, the thickness of the lead would give rise to seams which were somewhat bulky by comparison with copper, or even aluminium.

Successive lengths of copper can be joined by double-lap welts, or by drips. Welts must be formed with freedom for thermal movement, and must be clipped to the roof deck. In theory it is best to use drips when the slope is under 5°, but in practice welts are often used, even on the shallowest pitches. Because copper is used in thin gauges, the thickness of the welt is only a few millimetres, and does not greatly restrict drainage. For the highest-quality work, the deck is rebated at the welts, to give a cleaner line to the roof surface.

It is not good practice to lay copper directly on to the timber deck. Just as for lead and aluminium, an underlay of inodorous felt, or a modern equivalent, is recommended.

Weathered copper takes on a distinctive green patina, which many people consider to be its most attractive feature. The first stage in the creation of the patina is the formation of the thin oxide layer. The next stage depends on the environment; near the sea a layer of copper chloride will form, in industrial areas copper sulphate predominates. In either case the final patina will be the beautiful light green for which copper is so prized.

The patina does not arrive overnight. In marine conditions it takes about five years to develop; in polluted industrial atmospheres it is likely to take up to ten years, and in normal city conditions may take as much as 15 years. In clean rural areas, it is possible that the patina will take 30 years to develop. As copper is chosen for extreme durability, its specifiers are prepared to wait for it to mature!

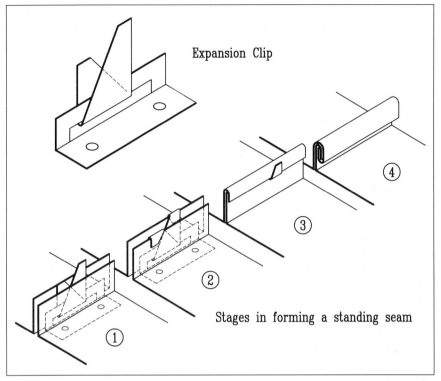

Expansion Clip

④

③

②

Stages in forming a standing seam

①

Figure 5.6 A method of anchoring a standing seam without restricting thermal movement.

The chemistry of the formation of the patina includes reactions with water. Walls drain too rapidly for such reactions to take place, so copper sheets on walls usually develop to a dull brown only; the green patina never appears.

Copper is the preferred metal for cladding domes. Its strength and hardness allow it to smooth out any small irregularities in the supporting structure, and its neat joints are unobtrusive and do not detract from the aesthetic appeal of the dome itself.

The London Planetarium is a well-known landmark. Its magnificent dome was constructed in 1958, and clad in copper. The copper is laid over a double layer of felt, on 25mm timber battens; most of the joints are standing seams, but every sixth joint is a batten roll. The roof has proved remarkably trouble-free.

(iv) *Zinc* has been available in commercial quantities for over 300 years, and can used in roofing, as a hard metal, in a similar manner to copper. However, despite its popularity in France, Belgium and Holland, it has never been widely used in Britain.

The failure of zinc to capture a significant share of the British market is probably due to its slightly poorer performance in terms of durability. In tests at worldwide locations, zinc was consistently out performed by aluminium, copper and lead. In terms of pitting depth, or metal lost, zinc was corroded at more than double the rate of those other metals.

This is not to imply that zinc is unsuitable for roofing purposes, but that the other metals set a standard of excellence which is difficult to match. A decision to use zinc could be based on price, and the required life of the roof. The MAFF Map, described in Chapter 4, is a good guide as to those parts of the country where zinc roofing would corrode rapidly, and those where it should prove durable.

Zinc roofing is formed with either batten rolls or standing seams, and with welts or drips across the slope. In fact, zinc is used in a very similar manner to copper; but in thicker gauges. It is possible to space the batten rolls at up to 890mm apart.

(v) *Galvanized steel* is only suited to very short-term applications, unless it is protected with an organic coating. With a coating such as plastisol, it may be expected to provide a reasonable service life - perhaps 20–30 years with adequate maintenance.

Like zinc, galvanized steel is used quite extensively in other parts of Northern Europe, but thus far is not widely used in Britain. However, it is relatively inexpensive, and its low rate of expansion enables it to be used in long sheets. It may therefore be anticipated that this material is likely to be increasingly applied where low initial cost outweighs long life requirements.

It is recommended that galvanized steel is not used at very shallow pitches, say under about 4°. It is important that water should be drained as thoroughly as possible; standing water will seek out weak points in the coating system, and set up local corrosion. If steel is allowed to rust, it can be penetrated in an extremely short time.

With this proviso, it is possible to use galvanized steel like copper or zinc, i.e. treat it as a hard metal, and use batten rolls, standing seams and welts. Care must be taken during site working, or the coating may be damaged, with disastrous consequences for the long-term performance of the roof.

(vi) *Stainless steel* describes a family of steel alloys which may include quite large proportions of chromium, molybdenum and nickel. For roofing purposes the austenitic grades, that is grades with a minimum of 18% chromium and 10% nickel, should be used. In aggressive industrial or marine atmospheres, the addition of 2.5% molybdenum will increase the durability of the alloy.

This is a very modern material by comparison with the other metals. It was first used in the USA, and in Europe, about 60 years ago. The first roofs performed well, and there is every reason to believe that stainless steel is as durable as aluminium, copper and lead.

Stainless steel would normally be used as an alternative to copper. It is hard and may be formed into batten rolls, standing seams and welts. It is, however, a little more difficult to work than the traditional hard metals, and it is therefore best left to expert fixers, who have had appropriate training and experience.

Stainless steel can be supplied as 'terne-coated' metal; this means that the metal can have a thin coating of lead, very useful for providing a traditional appearance. Terne-coated stainless steel is popular for church roofing, as it is not subject to theft in the same way as lead.

To summarize, lead and high-purity aluminium are soft metals; copper, stainless steel and aluminium alloys are hard metals. These metals are all extremely durable; they are selected according to special local conditions, personal taste, aesthetic effect or to blend with existing buildings. Zinc and organic-coated galvanized steel do not offer such good durability, but may reduce the initial cost of the roof.

All these metals may be harmed by permanently wet conditions on the reverse side. Hard metals, however, do not require that the supporting deck must be fully continuous; it is possible to lay the metal (on inodorous felt) over boards with narrow gaps between. This makes for improved ventilation and allows for more rapid dispersal of any condensation.

The hard metals are all available in pre-formed sheets which create a pseudo-batten roll or standing seam effect. The range of possibilities is enormous, and the manufacturers of such systems will provide product data on request. Such standard systems reduce the need for skilled site operatives, but have the disadvantage that they are less adaptable to the special needs of a particular building.

Recent research has identified two hitherto unsuspected problem areas, however. These are the phenomena of wind pumping and thermal pumping.

Wind pumping is most easily explained by reference to Figure 5.7. When there is a high wind suction over a roof, the metal skin may be lifted, between its fixings, so that it is raised in the centre of the panel. This creates a void immediately below the metal, and this void must be filled with air, which is drawn in through the standing seams; if these seams are already holding water through capillary action, that water is drawn into the roof. If the water is trapped in the roof, it becomes a potential source of corrosion; it may be evaporated by heat, only to condense once more when the metal is cooled. The cycle can be repeated many times before the moisture is fully expelled through the seams once more.

Wind pumping does not usually occur on lead roofs, because the self weight of the lead resists most attempts by the wind to lift it. In contrast thermal pumping may occur on any metal roof, and was first identified in lead roofs.

Thermal pumping may occur when there is a sudden heavy shower on a sunny day. The metal roof is hot and, because the metal is a good conductor of heat, the substructure is also hot, as is any air held under the metal. The heavy rain bounces on the roof, and some is caught at the seams and held by capillary action. The metal meanwhile is cooled by the rain, in turn cooling the air immediately below the metal, and causing it to contract. To make up the original volume, more air must be drawn in through the seams, and this draws in any water already held there.

If the water is evaporated and then condenses again producing pure distilled water, in which lead is slowly dissolved. The problem is best avoided by good workmanship, adequate vapour barriers, porous underlays, and details which provide ventilation.

Another way to prevent thermal or wind pumping is to produce a roof without any air behind the metal skin, and under which there is no possibility of a void being formed. This can be achieved by bonding the metal to a suitable backing such as plywood or chipboard. This is not a task to be undertaken lightly; success depends on the quality and permanence of the bond, and on the panel size being tailored to the thermal expansion limitations of the metal.

The bonded panel will still fail if vapour can pass through to the metal and condense on its inner face. There are specialist manufacturers of these systems whose advice should be sought on the type of panel to be used and the method of application.

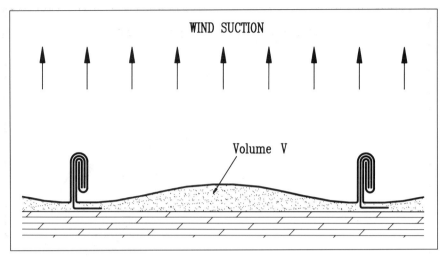

Figure 5.7 The mechanism for wind pumping.

Bonded panels can be invaluable where aesthetics have a very high priority. The factory-made panels achieve a degree of flatness which would be impossible to produce in a site application, and these panels are often specified for mansard slopes, and fascias.

STANDARDS
BS CP 143:   Part 5:1964 - Zinc roofing
               Part 12:1970 - Copper roofing
               Part 15:1973 - Aluminium roofing
BS 6915:1988 - Lead roofing

FURTHER READING
Lead sheet in building;a guide to good practice - The Lead
     Development Association.

INDEX TO ADVERTISERS

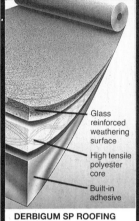

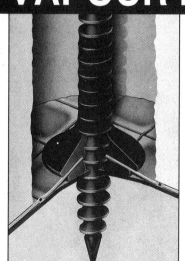

FULLY SUPPORTED MEMBRANES   6

Plate 6.1 Ballast over non-absorbent insulation is located above the membrane in this inverted roof. (By courtesy of Euroroof Ltd.)

This chapter, like Chapters 3, 4 and 5, is concerned with the weather skin, which in this case is a continuous, fully supported membrane.

Unlike the weather skins described in the previous chapters, membranes are not directional. They do not have side laps, seams or ribs and the typical appearance of a membrane is an unbroken surface with a relatively smooth finish. This may be a mineral finish, or a covering of stone chippings.

Membranes may be applied as a liquid which sets in place, rather like a thick paint. They may also be applied as thicker compounds such as asphalt, or as multiple layers of felt, or as modern single-layer systems.

An unobstructed surface provides greater safety for foot traffic. In fact, membrane systems are easily the most popular choice for any applications where regular access is anticipated. The finished roof has a continuous covering which is effectively joint-free; in theory the roof could be completely flat, but for reasons which will be discussed later, a minimum pitch of around 1° is usually specified.

Because the weather skin is smooth and continuous, membranes are particularly suitable for roofs with complicated details. These could include multiple slopes with numerous hip and valley intersects, or curves such as barrel vault roofs. Membranes are also regularly chosen for roofs with many openings (e.g. for rooflights or ventilators).

It is tempting to make a sweeping generalization that, just as tiles are a relatively heavy roof covering for steep pitches and small spans, so membranes are relatively light roof coverings for shallow pitches and wide spans. This overlooks the many domestic applications for membranes, such as coverings for sun-lounge extensions and garages; however, it is a reasonable statement in commercial and industrial applications.

Table 6.1 summarizes the common types of membrane. The systems will be discussed in greater detail, later in this chapter.

Asphalt is the oldest of these membranes, occurring in nature as limestone impregnated with bitumen. In the time of King Nebuchadnezzar, Babylonian architects virtually rebuilt the city of Babylon, using asphalt extensively on buildings, roads, bridges and sewers. It is known that Christopher Columbus caulked his galleons with bitumen from the natural deposits in Trinidad. The Trinidad asphalt lake is still the most abundant source of natural asphalt.

About 200 years ago a Swedish designer produced a composite roofing material by coating paper with tar, but it took another hundred years to develop this idea into anything that we would recognize as roofing felt. The first roofing felts were made

Table 6.1 A comparison of the four most common membranes.

Membrane	Weight Kg/m^2	Description
Liquid	1–2	The liquid must be applied to a stable, dust free surface. These systems are useful for repairs and maintenance work.
Asphalt	40–60	Durable, hard wearing surface. Suitable for foot traffic when surfaced with a "hard" grade.
Multi-layer	15–20	Many felt types available; it is vitally important to select felt an appropriate type for each layer. These are versatile systems.
Single-layer	2–5	Robust membranes, often formed from modern plastics; produced membranes with special surfaces. These are proprietary products, the manufacturer will advise upon their use.

N.B. The weights quoted are approximate only, and apply to the *membrane*. They do not include stone chippings or other ballast, nor is the weight of the structural deck or insulation included.

by saturating a fibrous sheet with bitumen, and the fibrous sheet was rag-based.

These felts were successful in the applications for which they were first designed. However between about 1960 and 1975, several changes occurred. The fashion in building moved towards shallow-pitched roofing for which membranes were most appropriate. At the same time developers were attempting to build wide, clear spans, with minimum obstruction of the floor space; this required light weight, which was achieved by means of multi-layer felt on lightweight metal decking. Such roofs move considerably as a result of snow loads, wind loads and thermal expansion. The felts of the day were unable to cope with these movements, and failed due to cracking or tearing.

During the same period there was a great increase in the use of thermal insulation. Until this was properly understood, many roofs suffered from severe condensation problems.

The premature failure of a large number of felt roofs turned many architects against such systems. This allowed concealed-fix metal systems to establish a foothold in the market. The felt manufacturers responded by developing improved products, and single-layer systems were also introduced. Today single-layer membranes, multi-layer felts and concealed-fix metal are able to co-exist in the same market place. Each is capable of providing a first-class roof, and the designer is able to specify, with confidence, the system most appropriate to his building.

It must be appreciated that these are fully supported systems, and the support can be provided in several ways. Figure 6.1 shows some of the more popular supporting decks.

Softwood boards on timber joists were the traditional deck for many years. Plywood and chipboard are acceptable alternatives to tongue and groove boards. This arrangement is often used for short-span domestic applications, where there is to be a ceiling at lower level with space for thermal insulation. It is seldom economic for large spans over factories or other commercial buildings.

Insulation board on metal decking is probably the most common supporting system. The membrane receives continuous support from the insulation board, but the board receives only intermittent support from the deck. This was very important when insulation levels were low and boards were thin; foot traffic could collapse the boards in the unsupported areas. Traditional decks were designed with this in mind, and the boards were never made to span more than 75mm. Today insulation levels are higher and boards are thicker; they are usually strong enough to cope with greater unsupported spans, and modern decks take advantage of this. However, it is always wise to check that the proposed deck is compatible with the proposed insulation board.

Metal decks are produced in galvanized mild steel, stainless steel and aluminium. Galvanized mild steel is usually preferred on grounds of cost, but the other materials may be specified for particularly humid or aggressive conditions (such as swimming pools and paper mills). Profiled metal decks are usually between 50 and 150mm deep, and made from metal 0.7 to 1.2mm thick; they are capable of spanning 6m or more between supports.

Modern membranes can accommodate some roof movement; nevertheless it is

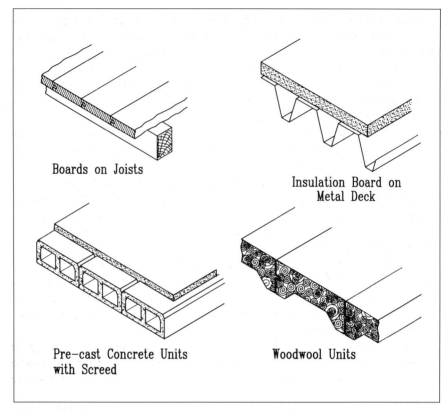

Boards on Joists

Insulation Board on
Metal Deck

Pre-cast Concrete Units
with Screed

Woodwool Units

Figure 6.1   Continuous decks in some popular materials.

sensible to restrict such movement. Current practice is to design the supporting deck
such that it will not deflect, under any load, by more than a two hundred and fiftieth
part of its span. For example, a deck spanning 2.5m should not deflect by more than
10mm under any combination of snow, wind, self weight and foot traffic. This
limitation of deflection is an effective restriction of movement within the weather
skin.

Pre-cast concrete units, or in-situ reinforced concrete, can provide a very stable
deck, and can accommodate spans up to 6m and beyond. However, the units are
expensive, and their weight increases the cost of the building structure. They are
sometimes chosen for their acoustic insulation properties, as mass is the main
ingredient for sound insulation. Hence they could contain the noise from machinery
near a residential area, or could improve working conditions in an office near an
airport.

A concrete roof deck is a good choice when there is to be regular foot traffic. In
such cases the membrane would usually be asphalt, and the falls for drainage would
be formed in the screed. Concrete decks are also useful in restricting the potential
spread of fire in high-risk applications, and can also be used when there is a possibility
of adding a storey to the building at some future time.

Woodwool units have characteristics which make them ideal for some applications. They can span up to about 6m and have sufficient weight to provide reasonable acoustic insulation, but they do not penalize the structure to the same extent as concrete. Their inner surface provides excellent acoustic absorption which reduces resonance in the building. This is highly desirable in sports halls, classrooms and gymnasia, and in multi-purpose buildings; woodwool roofing is also a popular choice for television studios, theatres and concert halls.

Woodwool can be supplied with various factory-applied finishes on its upper surface. These include felt, for temporary weather protection, and cement screed, for enhanced sound insulation.

Whatever type of deck is specified, it is usually necessary to provide some thermal insulation (see Chapter 8 for details of many insulation types). The membrane requires fairly rigid support, so some form of board is also needed. Polyurethane, polyisocyanurate and extruded polystyrene are all regular choices. Phenolic foam, mineral wool slab and cork board are also used on occasion.

Chapter 9 explains the way in which condensation occurs within an insulated roof construction, and it is emphasized that water vapour *must* be excluded from the roof construction. This is of paramount importance in membrane roofing, as moisture cannot escape through the membrane (in tiles, profiled sheets and supported metal, it is possible for vapour to be driven out at the overlap joints).

Vapour is excluded by means of a vapour barrier, or vapour check. A typical arrangement is shown in Figure 6.2. The importance of the vapour barrier cannot be over-emphasized. The failures of many membrane roofs during the 1970s were

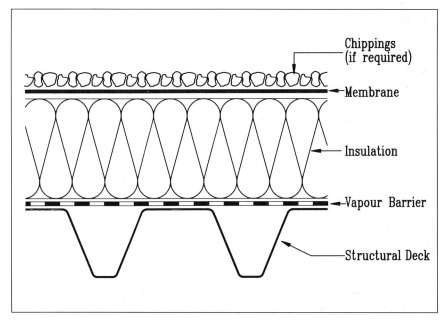

Figure 6.2   The usual components of a fully supported membrane roof.

attributable to inadequate vapour barriers. (Some proprietary systems involve the use of ventilation channels through the insulation, and ventilation openings through the membrane. These are beyond the scope of this book, but full details and recommendations are available from the manufacturers.)

The various systems are now described in more detail.

Asphalt is more properly described as 'mastic asphalt'. It consists of a mixture of inert aggregate, usually limestone, and bitumen. The limestone aggregate is graded, and the bitumen fills the voids in the same way that cement fills the voids between the aggregate in concrete.

It is most frequently applied over concrete, and the asphalt is normally separated from the concrete by means of a sheathing felt, which is loose-laid with nominal overlaps. It prevents the asphalt from becoming damaged through building movements, and also allows vapour to escape.

Roofing grades of mastic asphalt are covered by BS 6577 and BS 6925. It is usually mixed and graded under controlled conditions, and delivered to site in block form. For roofing purposes it is applied in two separate layers of about 10mm each. The layers can be of different grades, especially when regular foot traffic is envisaged.

The material is heated to about 250°C, when it can be spread and compacted. The work proceeds without joints, but if the area is too great to complete in a single day, a temporary stop can be formed against a timber former; the work can then be restarted the following day. At the restart, the edge of the existing asphalt must be thoroughly heated to create the most effective joint, and the upper layer must have any joints staggered so as to avoid a weakness penetrating the membrane.

At walls, parapets, or upstand kerbs, the asphalt is formed into skirtings at least 150mm high. The end result is a complete and continuous membrane following the shape and profile of the roof.

British Standard CP 144:Part 4 provides much valuable information on the use and application of mastic asphalts in roofing.

The best-known form of membrane roofing is the *multi-layer system*, which is also called *built-up roofing*. The membrane consists of several layers of felt, bonded together.

Roofing felts typically consist of a reinforcing fibrous base impregnated with bitumen. There are many different products and many manufacturers. The reference guide for bitumen-based roofing felts is British Standard BS 747. This standard uses a simple code of numbers and letters to identify the product. The fibrous base is represented by a single number, for example hessian fibre falls in type 1, asbestos in type 2, glass fibre in type 3, etc. The most successful modern felts have a non-woven polyester base, and these are included in type 5 in the standard.

The standard also uses a letter code to further describe the various products. 'B' implies a fine granule finish to the surface, 'E' represents mineral surfacing. Thus a type 3E felt would have a glass-fibre base and a mineral surface.

Great improvements in roofing felts were brought about by the introduction of polyester bases, but this was not the only source of improvement. Bitumen hardens with age and exposure to ultraviolet radiation, becomes brittle and the felt could then fail. The addition of synthetic rubbers to the bitumen increases its elasticity and improves its capacity to accommodate building movements.

The felts can be fixed to the roof in several ways, but the method most frequently adopted is adhesion by means of hot bitumen. The bitumen is heated to around 200°–250°C, and spread uniformly, it sets into a waterproof layer with good bond strength. This operation should not be carried out in damp or frosty conditions.

Figure 6.3 shows how a roof may be constructed, using different types of felt at the various levels. A sheathing felt used as the vapour barrier is bonded in hot bitumen to make it adhere to the deck, and to seal its overlaps. The insulation board is bonded, in hot bitumen, to the sheathing felt. The outer layers of the roof will experience large temperature changes throughout the year, causing expansion and contraction which will attempt to delaminate the membrane from the insulation. These thermal movements may be more easily accommodated if the membrane does not adhere continuously to the insulation. A discontinuous adhesion is achieved by a perforated felt laid dry over the insulation, with hot bitumen then spread over it. Adhesion of the perforated felt to the insulation will only occur at the perforations, and the next layer of felt will adhere continuously.

The perforated layer is the correct choice when the insulation is polyurethane, or when fixing to plywood or concrete; a continuous felt would be preferable when laying over mineral wool slabs, cork or fibreboard.

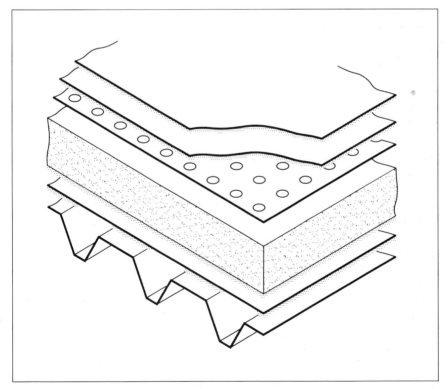

Figure 6.3   The various layers in a fully bonded roof.

The top layer, or cap sheet, is usually mineral-surfaced to protect the bitumen from direct exposure to ultraviolet radiation, and to provide a more durable finish. It is also possible to lay a covering of stone chippings. These protect the bitumen from direct light and from heat build-up, but the stone chippings are relatively heavy and are not very popular with structural designers.

When laying layers of felt, setting out should ensure that joints do not coincide in successive layers. Staggering the joints is better, and running joints at right angles, between successive layers, is better still.

There are other, less common methods of fixing the roofing felt. Over timber or plywood, the underlay may be nailed, using galvanized nails with large heads; this gives an intermittent fixing similar to that achieved with perforated underlay. Subsequent layers may be hot bitumen-bonded as before.

Cold adhesive is an alternative to hot bitumen, and various proprietary adhesives are available. These are not common on major projects, but can be extremely useful for small works, and for the do-it-yourself roofer.

Another possibility is the use of 'torch-on' felts. These felts are supplied with a coating of bitumen, which is sufficiently thick to provide the adhesive layer. The felt is rolled on to the roof slowly, and the bitumen coating is melted with a butane torch, just before it contacts the deck or underlay. This method removes the need for a bitumen boiler on site, and it is particularly useful for small works or projects where roofing is not a continuous operation.

Depending on the relative position of the insulation, roofs can be classified as 'warm', 'cold' or 'inverted' (see Chapter 8). The roof construction shown in Figure 6.3 is a 'warm' roof. Membranes can be used for each of these roof types. Warm roofs are the popular choice for industrial applications where the appearance of the deck is not of great importance. Cold roofs are often used in domestic situations; the membrane is laid on boards or plywood, and the insulation is at lower level, directly above the ceiling. Light fittings and other services can seriously reduce the effectiveness of the vapour barrier.

An inverted roof has the insulation over the membrane. The insulation must not become waterlogged, or be degraded by heat, cold, damp, light, etc. Extruded polystyrene and dense mineral wool are the usual choices. The insulation is ballasted by stone chippings, or sometimes by paving slabs, which protects the membrane from thermal movement, thermal shock, ultraviolet radiation, impact and foot traffic. The disadvantage with such roof systems is that leaks can be terribly difficult to trace and to cure.

Figure 6.4 shows the way in which a common fault in felt roofing may be avoided. A roof structural joint may open and close, by very small amounts, even when it is not a true expansion joint. Suppose a joint is 0.5mm wide, and that under some loading condition it opens to a full 1mm wide. The actual amount of movement is only 0.5mm, but the increase in the joint width is 100%.

If the felt adheres tightly to the deck at either side of the joint, it too will need to stretch by 100%. Not all felts can accommodate such movements, particularly as they become older; this can lead to splitting or tearing of the felt, and failure of the roof. The solution indicated in Figure 6.4 is to lay a narrow strip of felt on the deck,

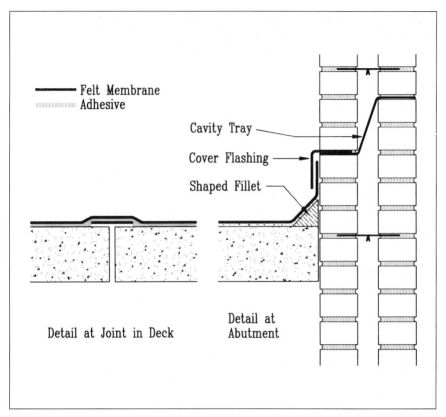

Felt Membrane
Adhesive

Cavity Tray

Cover Flashing

Shaped Fillet

Detail at Joint in Deck

Detail at Abutment

Figure 6.4   Two ways of preventing damage from thermal movement.

covering the joint. This strip does not adhere to the deck. The felt finishes are bonded to the deck and to the felt strip. If the strip is 50mm wide and if movement occurs, as described above, the felt is required to stretch 0.5mm in 50mm, only 1% instead of 100%. Such a small relative movement is most unlikely to cause splitting.

Felts do not readily withstand being folded through 90°. Such treatment will often lead to cracking. One solution is to use shaped fillets of timber (or of dense insulation) at any upstands or abutments. The arrangement is shown in Figure 6.4. If the deck can move independently of the wall, the felt should not adhere to both; the upstand may be trapped behind a cover flashing to maintain the weather protection without stressing the membrane. The upstand should be at least 150mm high to give proper weather protection.

There are a large number of *single-membrane systems*. These are proprietary products, and are not included in any current British Standard, although the better known systems have Agrément Certificates.

As the name suggests, the important feature of these products is that weather protection is achieved by means of a single membrane, as opposed to the multiple plies in a built-up system. Various materials are possible, but most systems are based on reinforced PVC, and the reinforcing fibres may be glass or polyester.

In the case of smaller roofs, the membrane can be welded in a factory and delivered to site as a single piece complete with formed edges for eaves and verges (this is rather like a scaled-up fitted bedspread!). The membrane can then be ballasted with stone chippings to make a continuous membrane without any direct fixings or adhesive.

For larger roofs, the membrane is delivered to site in wide rolls which are spread on the roof. Successive strips are jointed by welding, by heat or chemical reaction, depending on the particular system.

Mechanical fixing is the usual arrangement. Figure 6.5 shows two possible fixing systems, using self-tapping screws or special rivets. The supporting deck may be only 0.7mm gauge steel, which does not offer great anchorage for screws, so it is important

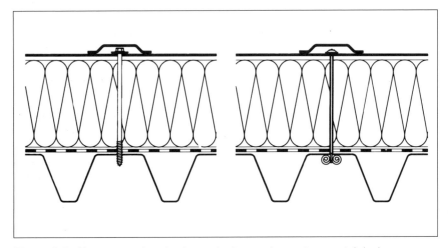

Figure 6.5   Two ways of anchoring a single membrane to a metal deck.

to select the right type of screws, and to use them in sufficiently large numbers. Aluminium decks are often thicker than steel (0.9 and 1.2mm are the usual gauges). However, these gauges are not sufficient to offer safe anchorage for self-tapping screws. There would be a danger of overtightening causing the threads to strip and thus weakening the fixing still further. For aluminium decks is it best to use special rivets which open into wide 'petals' or 'claws'.

The fasteners may be fitted with large spreader washers, or may be applied through some sort of continuous strip which acts as a load spreader. The exact arrangement will be a design requirement of the system. Whichever method is used, the fastener heads and load spreaders are covered with a strip of the membrane material. This strip is welded to the roof by the same process as that used to weld the seams. The fasteners are thus totally protected from the weather.

Some systems use fasteners at the edge of the sheet only. The next sheet covers the fasteners and is welded to the previous sheet to form a weathertight seam which protects the fasteners. This is an attractive concept, but has certain limitations. If wide sheets are used, the spacing of the fasteners becomes large; this may not be satisfactory if wind uplift loads are high. The alternative is to use narrow sheets, but then the number of seams is greater, and the risk of leaks is increased.

The fasteners anchor the membrane to the deck. If there is a vapour barrier it will inevitably be penetrated by each fastener. Unless precautions are taken, the vapour barrier will lose much of its value as a result of these penetrations.

Vapour barriers are not all equally effective. Bitumen-based felts are likely to outperform polythene in this respect. Manufacturers and suppliers may be able to advise on which materials will give the best results in a given situation.

The type of fastener is also influential. Certain self-tapping screws can be supplied with a flexible, adhesive coating. The action of tightening the screw causes the coating to spread and bond to the vapour barrier. These fasteners may provide a complete solution in some cases, but it is always important to seek the manufacturer's advice on the correct installation techniques.

The colour of a roof surface can have a profound effect upon its performance. Because pale-colours reflect the sun's rays, whereas dark colours absorb them, dark-coloured roofs can be up to 20°C hotter than otherwise identical roofs in a pale colour. Most single membranes are supplied in pale colours to avoid the damaging consequences of high temperatures.

Single-membrane systems are relatively new in Britain, although their use has increased quite dramatically since about 1980. Some systems, developed in mainland Europe, have been proved in the field for sufficient time to inspire confidence. There seems little doubt that single-membrane systems, properly installed, can provide a service life in excess of 20 years, thus matching the life expectancy of a well constructed multi-layer system.

Liquid roof coatings are economical and relatively simple to apply, but do not offer the same durability as the other systems discussed in this chapter. They can be very useful in carrying out repairs, and appeal to do-it-yourself enthusiasts.

A typical product would be bitumen-based. The bitumen may be emulsified for cold application, or modified by the addition of synthetic rubber. Most coatings of this type are fibre-filled for reinforcement and sometimes they are used in conjunction with a reinforcing glass scrim.

These coatings may be used over a wide variety of substrates including concrete, galvanized steel, fibre cement, lead, slate and tile. The surface must be thoroughly cleaned immediately before application of the coating, and any loose or powdery deposits must be removed. Coating must always be on to a dry surface.

The coatings may be applied by brush or spray. Spray application is specialized and may require a specialist contractor. A liquid-applied coating may be the ideal solution when there is a requirement to achieve a limited increase in the life of a large roof without going to the expense of a complete replacement.

Membrane roofs may be drained in different ways. Of course it is possible to drain into external gutters and then the gutter and pipe details will be the same as for any other roofing type. However, as the membrane should be continuous and waterproof, this offers the possibility of other approaches.

Sometimes no provision is made other than locating outlets to rainwater pipes at the lowest positions in the roof surface. All rainwater will eventually drain to the lowest points and hence escape through the outlets. This makes for simple construction but, during heavy rain, very large ponds may form on the roof; under such conditions water can seek out weaknesses in the membrane and cause roof leaks. Also, uneven settlement may result in ponding.

Discharge through rainwater outlets is more efficient when there is a reasonable head of water (see Chapter 17), a wide, shallow pond drains more slowly.

Some designers circumvent these hazards by incorporating deep channels into the roofs. All storm water enters these channels, and the risk of ponding on the roof slopes is removed. This is good for the membrane but introduces complications in the deck and the supporting structure.

A compromise arrangement is shown in Figure 6.6. The insulation thickness is halved under the drainage run (the 'gutter'). This creates a shallow channel which will cope with most of the rainfall, but which will overflow on to the adjoining roof during severe storms. The depth of water at the outlets will be greater, and this will increase rates of drainage.

This configuration is achieved without any modification of the deck or its supports. Reducing the insulation thickness locally has the added benefit that heat escaping from the building is likely to keep the main drainage runs clear of ice. This could be very important in the event of a sudden thaw following a heavy snowfall.

Fully supported membranes are unique roofing products in that they provide a complete weatherproof skin over the whole building. They can thus provide a degree of security which is impossible to achieve with any other roof covering. However, because they have no joints through which they may 'breathe', they can suffer from condensation. The design process should identify any potential condensation risks, and include strategies for their removal.

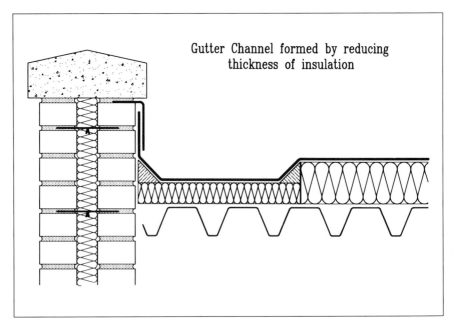

Gutter Channel formed by reducing thickness of insulation

Figure 6.6 Membranes can accommodate defined drainage runs.

STANDARDS

BS CP 144:	Part 3:1970 - Built-up bitumen felt roof covering.
	Part 4:1970 - Mastic asphalt roof covering.
BS 747:1977 -	Specification for roofing felts.
BS 6229:1982 -	Code of practice for flat roofs with continuously supported coverings.
BS 6577:1985 -	Specification for mastic asphalt for building (natural rock asphalt aggregate).
BS 6925:1988 -	Specification for mastic asphalt for building and civil engineering (limestone aggregate).

FURTHER READING

Roofing Handbook - The Flat Roofing Contractors Advisory Board

7

LINING SYSTEMS

Every roof has two surfaces: the outer one, which is exposed to the weather, and the inner, one which may be a visible feature of the interior of the building. Chapters 3-6 were all concerned with the external surface, and the ways in which different materials and products are used to resist the weather. This chapter concentrates on the inner surface, or *roof lining*.

In most domestic housing applications, there is a ceiling which is divorced from the roofing. This type of construction, best described in books on building, is not roof lining within the meaning of this chapter. The lining systems to be discussed here are those which form an integral part of the roof construction, and are on view to the occupants of the building. Their aesthetic qualities will be taken into account by the architect or specifier. It follows that much of this chapter will be about commercial buildings such as factories, warehouses, offices and supermarkets, but it will also apply to sports halls, swimming pools, ice rinks and other centres of entertainment.

Linings may be fixed *under-purlin* or *over-purlin* (Figure 7.1). Under-purlin lining is attractive in that it provides a plain, uninterrupted surface, and the purlins are hidden from view. However, it is more difficult to fix as it requires access to the underside of the roof. This is best achieved with a travelling tower scaffold, but this

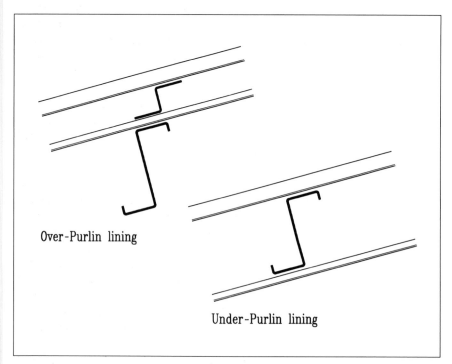

Over-Purlin lining

Under-Purlin lining

Figure 7.1 Linings can be located either over or under the purlins.

cannot be used until the floor slab is completed. In any case, the fixing of under purlin-lining is likely to be physically demanding, overhead work.

By comparison, over-purlin lining is easy to install. The materials are stacked on the purlins and placed in position prior to fixing. The state of the floor slab is thus of no consequence, and the roofing can proceed regardless of other items in the building programme. In fact, the ability to complete the roof at an early stage can shorten the programme for other building works by providing a protected environment.

Some years ago, under-purlin lining was fashionable for all prestige applications, but it gradually fell out of favour because of its high cost (for reasons given above), and because of detail difficulties at the rafters or trusses. More recently, designers have found ways to compensate for the less attractive appearance of over-purlin lining; for example, purlins may be made from RHS sections and painted in contrasting colours, and the cost of this is usually much less than the extra cost of under-purlin lining.

Today there are three main reasons for specifying under-purlin lining. First, the architect may wish to achieve a special aesthetic effect which requires that the purlins are hidden. Second, the building may require regular cleaning and disinfecting (e.g. food processing premises), purlins can become dust ledges and under-purlin lining can improve hygiene. Third, roof insulation levels may need to be increased; if the existing roof is otherwise sound, under-purlin lining may provide the simplest means of carrying out the work.

The lining can perform a number of functions. It can enhance the appearance of the roof, support the insulation (or provide the insulation), improve the acoustic properties of the roof, support the wind and snow loads, improve fire resistance, impede the passage of vapour and conceal plumbing and electrical services. Some of these functions are discussed in this chapter, others are covered elsewhere in this book.

Whenever the lining is on view (this includes most commercial applications, and many others), its appearance is important. The designer will seek to achieve the most attractive finish possible within his budget. The appearance of the lining is influenced by its profile, colour, gloss level, material and texture. In this context, it is interesting to review changing fashions over the last 30 years.

During the 1960s, most factories and warehouses were roofed with a single skin of asbestos cement, and the inside of the roof was therefore the pale grey of natural asbestos cement. It followed that, when insulation became more widely used, an asbestos cement lining sheet was introduced. Soon, plasterboard was seen to provide a cheaper alternative, with a flat, white surface on the underside. Later, insulation boards replaced the plasterboards. These also had a white underside, and also simplified the construction, but were not totally trouble-free systems, and profiled metal linings (in conjunction with mineral fibre insulation) became popular. Composite panels have steadily increased their market share. These have metal liners which can have very flat profiles because their strength is derived from the composite action of the panels.

All these systems are in current use to some extent. Figure 7.2 shows the most common arrangements (insulation boards can be fixed in T-bars, in the same way as plasterboard). Shallow profiled metal, usually in thin gauge steel with a white enamel finish is most popular today. Shallow profile sheets are also available in natural mill-finish aluminium, and these remain bright for many years in the internal environment.

Bright aluminium is not widely accepted in Britain, although it is quite popular on the continent. Steel is little less expensive initially, but aluminium provides exceptional durability. It has been argued that condensation within the roof construction could lead to corrosion of steel; supporters of this viewpoint maintain that aluminium should be preferred. In fact, this is a dubious claim; the lining sheets are on the warm side of the insulation, and should not be subjected to regular wetting.

However, if there are leaks in the outer skin, it is possible for large quantities of water to flow over the lining sheets. Roof details are usually designed so that lining sheets drain to the gutters, and it is therefore possible that roof leaks may go undetected for long periods. Regular wetting of steel lining sheets could certainly lead to corrosion, but there are no widespread reports of such problems. It is possible that the roof pitch allows the water to drain readily, and the heat of the building dries up the residue, but it is also possible that steel linings have not been in use long enough for corrosion problems to develop. Fashions in roofing have changed rapidly

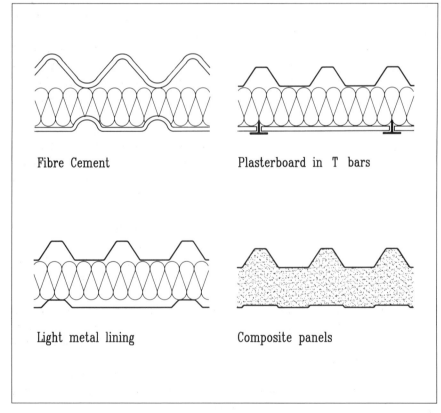

Figure 7.2 Some of the most popular lining systems.

in recent years, and it may take some time before the strengths and weaknesses of some current products are fully appreciated.

The structural requirements of the lining vary considerably between different systems. The lining will usually need to support its own weight between structural members, e.g. as a profiled sheet spanning between purlins, and it may also have to support the weight of the insulation, particularly when this takes the form of a quilt. It is also possible that the lining may be the main structural element (e.g. roof decks), and may be required to support all applied roof loads, including snow, wind and foot traffic.

One loading case which is not always fully understood is that of internal pressure or suction. This condition is described in Chapter 16, and results when a door or window allows wind gusts to enter the building. These loads can be quite large, and it is not unknown for plasterboards, or insulation boards, to be dragged out of their T-bar supports by unexpectedly severe wind loads.

In the case of roof decks, all the load is supported by the deck; internal wind loads are simply one component of the overall design case. Similarly, composite panels act as a single structural element.

When boards are placed in T-bars, it is the T-bars which provide the structural strength to span between the purlins. The boards simply span the shorter distance between the T-bars. The T-bars must therefore be strong enough for the task, and they should not be viewed as trims to cover the joints in the boards.

Another form of loading to be taken into account arises from the use or abuse of the building. It is possible for the lining to be subjected to impact, for example in a sports hall during football practice. This may demand the use of a robust deck formed from steel, thick gauge aluminium, concrete or woodwool. Plasterboards and thin-gauge metal are unlikely to be successful.

It is seldom necessary to take account of thermal movements in the design of lining systems. The internal temperature of most buildings varies very little throughout the year as the insulation retains the heat in winter, and keeps out solar heat in summer. However, designers should remain aware of thermal movement; they should also consider, for each building, whether there is a possibility of large variations of internal temperature.

Figure 7.3 shows three structural arrangements involving the use of metal liners. The lightweight metal lining supports its own weight, and that of the insulation; it also supports internal wind effects. The main roof loads are transmitted through rigid spacers to packs (usually plastic ferrules) located in the troughs of the liner. The packs are tall enough to hold the spacers clear of the liner profile.

The structural deck supports all the design loads. The external snow loads are transmitted through the rigid insulation. The wind loads may require the use of mechanical fasteners between the membrane and the deck, but such fasteners would be anchored in the deck itself.

Structural lining trays provide good structural support and a flat soffit. This can create a very attractive aesthetic effect. The version shown in Figure 7.3 carries spacers which, in turn, support the external sheet.

The same principle can be employed for tiled roofs, the spacers being replaced by

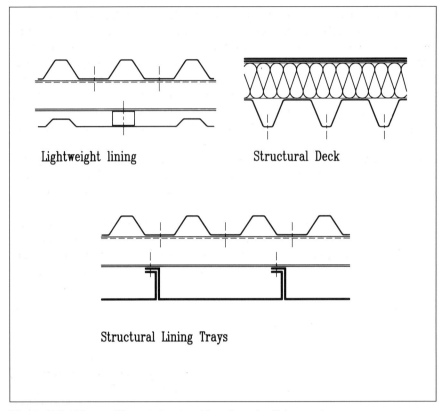

Figure 7.3   Three different structural functions for lining systems.

tiling battens. This form of construction has increased rapidly in recent years, and is claimed to reduce construction time and increase security (the tray becomes the sarking).

Special deep trays can be made to span between trusses or rafters, and this can remove the need for purlins; in effect, the edges of the trays become the purlins. It is also possible to make shallow, thin-gauge trays; these act in the same way as a lightweight lining, but provide a flat soffit.

Thus far, no mention has been made of construction loading. From the point of view of both speed and safety it is desirable that the lining system should be capable of supporting the weight of the roof fixers. In fact, this is seldom the case; lightweight systems, fibre cement and boards in T-bars, are easily damaged and must be treated with care. It is found that increasing the gauge of the lining, to support foot traffic, is prohibitive in terms of cost. Extra labour costs arise from extensive use of walking boards and roof ladders, but these are accepted as the lesser burden.

Where it is essential that the lining should support foot traffic, a cost comparison should be made between structural decks and reinforced lining systems. Other possibilities include rigid insulation boards, which spread the loads over a greater area, and composite panels which are normally walkable as soon as they are fixed.

All buildings have an acoustic performance; this is discussed in Chapter 14. In theatres, concert halls, debating chambers and classrooms, the acoustics may be vitally important. For many factories and warehouses the acoustics are of no importance whatsoever. However, there are a number of buildings where it is considered sensible to design for a level of acoustic performancebetween these extremes. For example sports halls, gymnasia, swimming pools and ice rinks, where it may be necessary to issue instructions to large groups, or supermarkets,where shoppers need to understand loudspeaker announcements, and for background music to be of benefit.

The roof represents a large area, and this can be 'tuned' to produce the required acoustic effect. Hard, smooth surfaces create echo effects; soft or irregular surfaces have a deadening effect. In large buildings the amount of echo often needs to be reduced.

Woodwool has an open-textured surface which has good sound absorption characteristics. It is a popular lining choice for sports halls as it is also robust.

Sound absorption can also be achieved by the use of perforated linings. Sound is allowed to escape through the perforations to be absorbed in the soft quilt insulation.

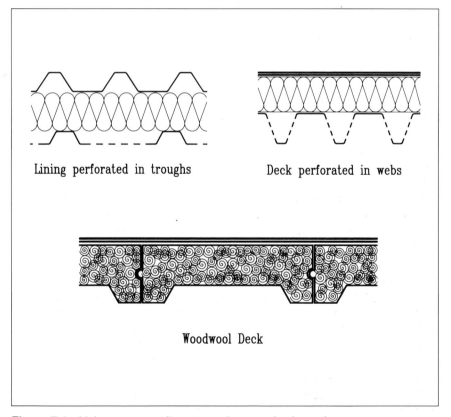

Lining perforated in troughs   Deck perforated in webs

Woodwool Deck

Figure 7.4   Linings can contribute towards acoustic absorption.

Most decks, trays and profiled sheets can be perforated, but there are certain guidelines as to the best approach.

It is generally agreed that the actual area of open space (i.e. the area of the perforations) need not be large; around 15%–25% of the roof area is normally found to be adequate. The perforations should be distributed as widely as possible, and many small perforations are preferred to a few large ones of equal area.

Perforations are usually in the range 2–5mm diameter, and are often spaced at about twice the diameter of the holes, on an isometric grid. It is often impractical to perforate metal sheets uniformly. Perforating weakens the profile considerably, and can also cause production problems. When metal coil passes through a roll former there is inevitably a small degree of 'wander', usually of just a few millimetres, which may be detected as small changes in the width of the overlapping edge. Wander of lines of perforations can be most unsightly when the line is close to a profile corner. Even if this could be overcome, it would be unwise to allow a line of perforations to coincide with a profile corner, as this could lead to cracking during manufacture.

For these reasons the coil is normally perforated in strips, so that the profile is not weakened in critical areas. Figure 7.4 shows some of the possibilities.

Lightweight liners rely on their ribs for strength. The ribs are spaced as widely as possible for maximum economy of metal; as the troughs contribute very little to the strength of the sheet, they may be perforated without serious structural consequences.

Decks are usually quite deep; this is to increase their stiffness, or resistance to deflection. As a result the webs are often far stronger than necessary. In such cases it is logical to perforate the webs.

Perforating metal linings for acoustic benefit can, however, also introduce certain less desirable influences. The actual process of perforating is carried out on the finished coil, prior to roll-forming. In the case of coated galvanised steel, the perforations are made after the coil has been galvanized and coated. Each perforation will expose unprotected steel around its perimeter, and the total area of unprotected steel can be very large indeed. If the conditions within the building are likely to be aggressive, then perforated steel may not be a wise choice. It may be better to use perforated aluminium, or to use a different type of material such as woodwool.

The perforations are only of benefit when they are backed by a soft material to absorb the sound. Typically this material is mineral-fibre insulation. However, there is an obvious risk that fibres could become detached from the insulation, and fall through a perforation into the building, thus entering the environment of the occupants. Depending on the nature of the fibre, it is possible that this could cause skin irritation or provoke attacks of asthma or other respiratory problems. This can be prevented by the use of a continuous vapour barrier, but the vapour barrier may reduce the effectiveness of the sound absorption; it is possible that increasing the diameter of the holes would compensate for a reduction in the degree of absorption. Sometimes it is necessary to carry out tests to establish the best combination.

Another possible acoustic lining system would be to use strips of timber with narrow gaps between. Gaps of about 6mm between timber strips of roughly 30mm would be the right proportions. Again the timber would be backed by soft insulation.

Stained and polished hardwood, fitted under the purlins, might be a suitable lining for a chapel or a library.

Linings may be affected by legislation governing performance in fire. Performance in fire is discussed at length in Chapter 15, where the standards and regulations concerning the behaviour of linings are listed and described. Linings should not promote or support combustion, nor should they emit toxic fumes as a result of exposure to heat or flames. In many respects roofs and walls are expected to behave in a similar manner. However, this is not true in respect of fire; walls are usually required to contain fire, while roofs are often expected to vent smoke and fumes to atmosphere. Chapter 15 explains these points in detail.

Sometimes the lining is expected to act as a vapour barrier or vapour check. The whole question of vapour control is covered in Chapter 9. Here it is sufficient to say that a vapour barrier is intended to prevent vapour, or humid air, from entering the roof construction. The standard vapour barrier consists of plastic sheets or bituminous felt with bonded or welded joints; it is imperative that the vapour barrier is continuous and unbroken.

Metal sheets, and metal foils, are impervious to water vapour, and form superb vapour barriers, provided that all joints are properly sealed. There is a clear economic advantage in using the lining sheets as a vapour barrier, and in omitting the layer of plastic or felt; this is specified quite regularly as a competitive arrangement.

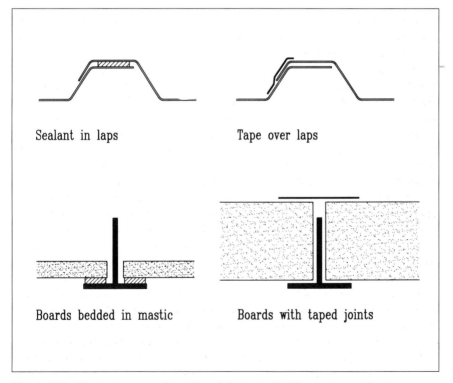

Sealant in laps   Tape over laps

Boards bedded in mastic   Boards with taped joints

Figure 7.5 Various ways of sealing linings against the passage of vapour.

While the concept is attractive, the actual results may be less pleasing; the seals must be continuous and unbroken, and this is very difficult to achieve in practice. Figure 7.5 shows some of the ways in which this is attempted.

The side and end laps of lightweight liners can be sealed with mastic sealer strips or gun-applied silicones. These will almost certainly be effective for much of the work. Unfortunately, there is no way of seeing whether the sealant is continuous within the lap, or whether it has bonded to both surfaces. The fixer cannot check his own work, and his supervisor cannot tell which laps contain sealant unless he watches the work continually (this comment also applies to the clerk of works). It is inevitable that there will be some weak or ineffective joints.

A slightly better method is to seal the joints by means of metallic adhesive tape. The tape is visible until the insulation is placed over it. The work can be inspected as it proceeds, and has a better chance of success. Some tapes do not adhere well in cold, wet or frosty conditions, and this can hold up progress, or lead to tapes being applied in unsuitable conditions. To be fully effective the tape must be pressed on to the metal, at both sides of the joint, continuously. As a large roof can contain several kilometres of adhesive tape, it is unlikely that it will all be properly applied.

Plasterboards may be bedded in mastic in the T-bars. This detail is often specified. Unfortunately, it less common to specify any sort of seal where the plasterboards join over a purlin; to be effective against vapour, all joints must be sealed.

Insulation boards, in T-bars, may have their joints sealed with adhesive tape. Here the seal is on the cold side of the insulation, and it is less concerned with resisting the passage of vapour than with preventing condensate from entering the building.

All four of the methods shown in Figure 7.5 also suffer from penetration by the fixing screws which hold the roof on to the building.

Various proprietary systems are available and these should be considered on merit for particular applications. The manufacturers will provide guidance on design details and installation.

Liners may provide adequate vapour resistance if carefully installed over low-risk areas. However, when vapour resistance is of paramount importance, it is usually better to include a separate vapour barrier. This is particularly important over swimming pools, ice rinks, paper mills and other high-humidity areas.

The lining may sometimes be simply a facing on the insulation. It could be a metal foil, a few microns thick, on a foam insulation board , or a thin sheet of metal on the underside of a composite panel, or bonded panel. Composite panels are efficient structural members, as the foam completely fills the panel and provides strength as well as thermal insulation. They are produced on dedicated machinery and are not available in a wide range of profiles. The user may accept what is on offer or seek an alternative roof construction.

Bonded panels are simpler to produce; within reason any outer sheet can be bonded to any insulation slab, and this to any liner sheet. This can be very attractive to the designer. However, the panel may not be perfectly filled, and voids may allow the occurrence of condensation. The adhesive is unlikely to provide as good a bond as is achieved by the foam in a composite panel, so the panel will probably not be as strong. Bonded panels may well be the only choice possible when a specifier insists on the outer sheet profile of his choice as well as a virtually flat metal liner.

Linings require very little by way of maintenance, and in some extreme cases they receive no attention whatsoever. Sometimes they may be dusted or washed annually. Old, yellowing coatings can be repainted, but in practice this very rarely happens. Rust stains on steel liners are probably evidence of roof leaks; unless action is taken promptly the lining will deteriorate to the point where it must be replaced.

Over-purlin linings can only be replaced after first removing the external roof covering. The designer should try to ensure that the lining will at least match the external skin for durability.

STANDARDS
The Building Regulations Part B2 - Internal Fire Spread (linings).
BS 476:Part 7:1987 - Method for classification of the surface spread of
    flame of products.

FURTHER READING
Trade literature.

THERMAL INSULATION

8

There are many insulating materials on the market, often supplied in several alternative forms. Roof construction can include insulation over the roof, under the roof, or within the roof; there may be a cavity above the insulation or below it! The range of options could become bewilderingly large. Within this chapter the common materials are discussed and compared; the more popular details are explained, and perhaps a few myths are exploded.

Insulation is available as quilt, slabs, boards, injected foams, beads and granules. Each of these has its place, and there is no single 'best' insulation. The strengths and weaknesses of each type will be discussed later in this chapter, but first it is necessary to explain the function of the insulation.

Deliberate thermal insulation of roofs is a very recent development in Britain. Thatch has long been used for roofing and is a good thermal insulator, but it became widely used because it was easy to apply, and used local materials; its insulating properties were a beneficial side-effect.

As recently as the 1950s, most buildings were constructed without thermal insulation. A requirement for insulation was first seen in the Thermal Insulation (Industrial Buildings) Act of 1957. The requirements of the Act were extremely modest, and could be met by about 14mm of mineral wool. Nevertheless, many buildings were exempt from even that specification!

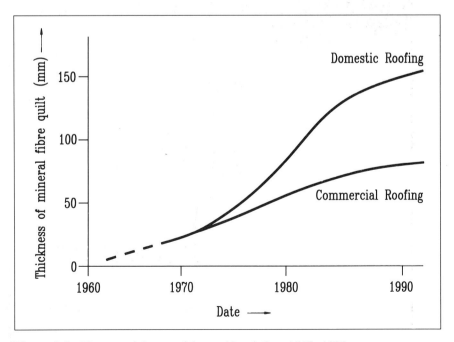

Figure 8.1   The growth in use of thermal insulation, 1960–1990.

For a further ten years or so, insulation continued to be seen as an expense rather than as a means towards more economic use of buildings. During the late 1960s and early 1970s several events had a major influence on the roofing industry. Industrial action, by coal miners and power supply workers, caused great increases in energy costs, and the escalation of oil prices added more inflationary pressure. Within a few years, energy costs increased out of all proportion to other costs. The immediate result was that building owners found that their heating costs were 'going through the roof ', because their heat was quite literally going through the roof.

It became attractive to conserve energy, and some specifiers started to demand thermal insulation. At the same time, government realized that conservation of energy would reduce the need to import oil, thus improving the balance of payments, and that it would also subdue demand for coal, thus helping its negotiations with the miners. The Building Regulations were amended to require thermal insulation in all heated buildings.

There have since been amendments to the Building Regulations to increase the amount of thermal insulation. The owners and designers of buildings have grown to appreciate the benefits of insulation, and trends are often ahead of the Building Regulations.

Figure 8.1 shows how typical insulation levels have changed over a 30-year period. The graphs simply illustrate what has occurred, and are not intended to portray any deeply significant scientific truths. Nevertheless, it appears that the trend is towards increasing levels of insulation, and we may assume that buildings in the future will have still more insulation than is used today.

In Figure 8.1 the comparison between levels of insulation is expressed in terms of thickness of mineral wool quilt. This was a useful way to demonstrate changes over a period of time, but it has limitations in other respects. Mineral wool is only one insulating material out of a vast number of possibilities. Some objective method is required to compare the relative insulating properties of the various materials.

To understand the method of comparison it is necessary to appreciate a little of the physics of heat transfer. When heat is applied to a part of an object, the heating effect travels through the object and warms parts which are not heated directly. For instance, a saucepan on a cooker is heated at the base, but the handle can become hot. This is called *thermal conduction*.

Conduction occurs because heat causes the molecules of the material to vibrate faster, and the molecules set up a vibration in their neighbours. The temperature of the material at any point is in proportion to the rate of vibration of the molecules at that point. It is important to realize that the molecules are not travelling through the material, they are simply vibrating in their own set place.

While this is a common behaviour for all materials, the extent to which molecules can affect their neighbours varies for different materials. When molecules have little effect on their neighbours, heat can only pass slowly; materials which greatly restrict the rate of heat flow are called *thermal insulators*.

In general, metals are good conductors, and non-metals are good insulators. The saucepan mentioned earlier as an example would usually be made of metal to allow

Table 8.1 Thermal conductivity and resistance values for a range of materials and roofing elements.

Material	λ (W/m °C)	Item	R m² °C/W
Phenolic foam board	0.018	External surface	0.045
Polyurethane foam board	0.022	Internal surface	0.105
Expanded polystyrene	0.033	Cavity (up to 100mm)	0.180
Mineral wool slab	0.033	Asphalt (per 20mm)	0.060
Mineral wool quilt	0.037	Vapour barrier	0.020
Cork board	0.042	Roofing felt	0.020
Softwood	0.130	(per layer inc. adhesive)	
Plywood	0.140		
Chipboard	0.150		
Plasterboard	0.159		
Fibre cement sheet	0.220		
Plaster/render	0.480		
Cement/sand screed	1.100		
Concrete tiles	1.150		
Reinforced concrete	1.440		
Clay tiles	1.840		

the heat to pass easily from the cooker to the contents of the saucepan. Its handle would be likely to be plastic to prevent its becoming too hot.

The property which governs the rate of heat flow through a material is its *thermal conductivity*. This is defined as the number of joules (J) which would pass through a sample in one second, given that the sample is 1m thick and has an area of 1m², and that there is a difference of 1°C between the two opposite faces.

For most applications, conductivity is written as λ (lambda) and is the number which is of most value for comparative purposes. It should be noted that joules per second are watts (W), so the units of conductivity become W/m °C.

Table 8.1 lists the popular insulation materials, together with many of the materials which may be incorporated into a roof construction and gives an accepted value of λ for each material. (There will inevitably be differences between batches of manufactured products, and even greater differences between batches of natural products. The tabulated values of λ are taken from British Standards, Trade Association Papers, Building Regulations, etc.)

The materials are listed in ascending order of conductivity. This is an arbitrary choice, done purely to highlight the best insulators.

The flow of heat is also disrupted by material surfaces, i.e. interfaces between the construction and the atmosphere. Here there is no material property or thickness to consider, so the units are not the same. It is usual to define the disruption to heat flow as *thermal resistance*, usually written as R. Strictly, R is measured in m² °C /W (see Fig. 8.2), but this has no significance except as an aid to calculation.

Some agreed standard values of R are also listed in Table 8.1, and these include thermal resistances for cavities and certain thin membranes. Every roof must at least have an outer and inner surface, so the insulation value will always be enhanced by surface resistances. The true resistance of a surface will depend on its texture, roughness and colour, but the tabulated figures are accepted values for use in calculations.

The conductivity and resistance values are useful only if they can be combined to determine the overall performance of the construction in resisting heat flow. This is done by first establishing the resistance of each element of the construction, and then combining these values. The resistance of a surface or cavity is taken direct from the table, the resistance of a layer of material is calculated by dividing its thickness, in metres, by its l value.

A sample calculation is included in Figure 8.2. The resistances are added together, and the reciprocal of their total is calculated. This gives a very important property of the construction, called the *thermal transmittance*, designated U. The 'U value' of a roof is the measure of its thermal efficiency.

The units of U are $W/m^2°C$. In other words, U is the number of watts (i.e. joules per second) which would pass through one square metre of roof, if the temperature outside was one degree lower than the temperature inside. Thus, U is a measure of the rate at which heat can escape; it follows that a small U value means a higher thermal efficiency than a large one.

In the example in Figure 8.2, a half cavity is included. When profiled sheets are laid over flat-topped insulation, there is a cavity under the crowns of the sheet, but not under the troughs; it is therefore considered appropriate to include half a cavity in the calculation.

The calculation in the example was repeated many times for different insulation and thicknesses, and for different insulation materials. The answers so obtained were the basis of the graphs included in the figure. The 'U value' is the important property, and this can be obtained by means of different thicknesses of the various materials.

Within the figure there is special emphasis on the values of $U = 0.45$ and $U = 0.25$. This is because these are the most important figures in the Building Regulations; 0.45 is the maximum value for commercial property, and 0.25 is the maximum for residential property (a value of 0.25 of course means more insulation than 0.45).

The Building Regulations are concerned with the performance of the building. They state standards which must be achieved, but allow the designer considerable freedom as to how this is to be done. It should never be assumed that the Building Regulations are a standard of excellence. In fact, they represent the minimum standard allowed under the law.

It may be illuminating here to consider a simple example. A factory has a roof area of 1000m², and the inside temperature is held at 18°C. On a spring day the outdoor temperature is 8°C. The factory roof meets the current Building Regulations as $U = 0.45W/m^2°C$

Figure 8.2 An illustration of the variation of thermal transmittance with thickness, for four popular insulants.

EXAMPLE CALCULATION

ELEMENT	t (m)	λ	R
Outer surface	--	--	0.045
Half cavity	--	--	0.090
Expanded polystyrene	0.075	0.033	2.273
Inner surface	--	--	0.105
		Total $\sum$ =	2.513

Where:—   t   = the thickness of the element in metres
λ   = the thermal conductivity of the element (W/m°C).
R   = the thermal resistance of the element (m²°C/W).

$$U = \frac{1}{\sum} = \frac{1}{2.513} = \underline{0.398 \text{ W/sq.m°C}}$$

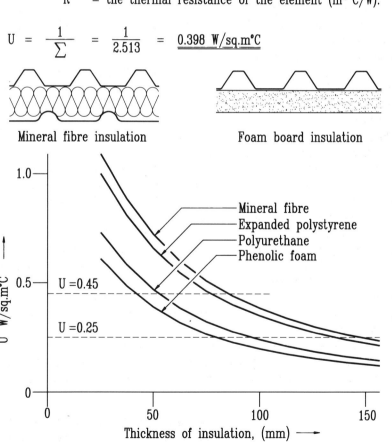

Mineral fibre insulation   Foam board insulation

The heat loss through the roof may be calculated as follows:

Heat loss = 1000 x 0.45 x (18 - 8) = 4500 W
$\qquad\qquad$ = 4.5 kW.

In 1965 a similar factory would probably have had a single-skin asbestos cement roof, with $U = 5.7$. The calculation would then have been as follows:

Heat loss = 1000 x 5.7 x (18 - 8) = 57 000 W
$\qquad\qquad$ = 57 kW.

This example shows very clearly just how great the changes have been in insulated roof construction. It is easy to imagine the 53 single-bar electric fires which would be needed to compensate for the additional heat losses through the old roof, on a relatively mild day. (The total heat losses for the building would also include the walls, floor, doors, windows, ventilation, etc., but that is beyond the scope of this book.)

Of course, the amount of heat entering a building, by way of the heating system, may exceed the heat loss. In such a case, the extra heat will raise the indoor temperature. This will increase the differential across the insulation, and this increases the heat loss. In due course a condition of equilibrium will prevail.

The method of calculating U, by combining resistivities, is so important that it is worth considering a more involved example. Figure 8.3 shows a proposed construction for the roof over a dormer, the calculation for U gives a figure of 0.247 W/ m^2 °C. Note the two layers of insulation; it would not have been possible to meet the domestic requirement ($U = 0.25$) with 100mm of mineral wool, and to increase the thickness would have increased the depth of construction. It is always an advantage to include some compressible insulation, so the arrangement sketched is a good compromise.

Each of the eight elements affects the value of U, but by far the biggest contributions come from the polyurethane and the mineral wool. Nevertheless, if the felt, chipboard or plasterboard were to be deleted, the construction would be unable to meet the requirements of the Building Regulations.

The method of calculation is well worth practising, and it can be applied to any combination of materials. Not only is it the calculation method to be used for new constructions, but it provides the means of analysis when an existing roof is to have its insulation increased.

The calculation is also the starting point in determining the variation of temperature through a construction. Figure 8.4 shows how the dormer roof calculation may be extended to provide the temperature at any point in the roof materials. It is assumed that the inside temperature is 20°C, and the outdoor temperature is -10°C. The heat flow through the construction is calculated as:

$\qquad$ Heat flow = 30 x 0.247 = 7.41 W/m^2.

The important point here is that this must be the heat flow through every element;

EXAMPLE ROOF BUILD-UP

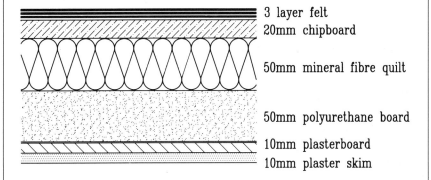

3 layer felt
20mm chipboard

50mm mineral fibre quilt

50mm polyurethane board

10mm plasterboard
10mm plaster skim

CALCULATION OF U VALUE

ELEMENT	t (m)	λ	R
Outer surface	--	--	0.045
3 layer felt	--	--	0.060
Chipboard	0.020	0.150	0.133
Mineral wool	0.050	0.037	1.351
Polyurethane	0.050	0.022	2.273
Plasterboard	0.010	0.159	0.063
Plaster skin	0.010	0.480	0.021
Inner surface	--	--	0.105
		Total $\sum$ =	4.051

$$U = \frac{1}{\sum} = \frac{1}{4.051} = \underline{0.247 \text{ W/sq.m}^\circ\text{C}}$$

Figure 8.3 The calculation of thermal transmittance for a multi-layer construction.

CALCULATION

ELEMENT	R	U_e	$\Delta T°$	$T_a°$
Outer surface	0.045	22.22	0.330	−9.67
3 layer felt	0.060	16.67	0.450	−9.22
Chipboard	0.133	7.520	0.980	−8.24
Mineral wool	1.351	0.740	10.01	1.770
Polyurethane	2.273	0.440	16.84	18.61
Plasterboard	0.063	15.87	0.460	19.07
Plaster skin	0.021	47.62	0.150	19.22
Inner surface	0.105	9.520	0.780	20.00
		Total $\Delta T°=$	30.00°	

U_e is the effective U value of the element, in isolation

$\Delta T°$ is the drop in temperature across the element

$T_a°$ is the actual temperature at the inner surface of the element

TEMPERATURE GRADIENT

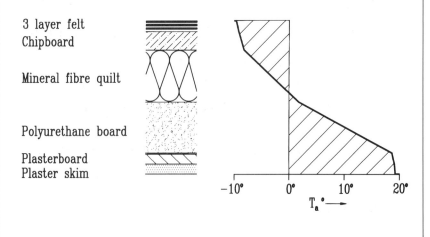

3 layer felt
Chipboard

Mineral fibre quilt

Polyurethane board

Plasterboard
Plaster skim

−10°   0°   10°   20°

$T_a°$ —→

Figure 8.4   The calculation of temperature variation through a multi-layer construction.

if the heat flows from inside the building to the outside, it must flow through each element on the way. Now the heat flow through an element is obtained by multiplying U by the drop in temperature, i.e. heat flow = U x temperature. This equation can be rewritten to give temperature, i.e. temperature = heat flow/U.

The value of U for each element is calculated by taking the reciprocal of its resistivity, designated U_e in the calculation. The temperature drop across each element is designated ΔT, and the temperature at the inner surface of each element is designated T_a.

Most of the temperature change is within the two layers of insulation, and there is a small change of about 1.5°C both above and below the insulation. So the felt and chipboard are very cold while the plasterboard and plaster skim remain warm. This calculation is not of any great significance in respect of the insulation, but will become vitally important in the consideration of condensation and vapour control, in Chapter 9. This is another type of calculation which merits practice.

Before leaving the calculation of U , there is one further case which should be considered. This is the case of insulation which varies in thickness in such a way that U cannot be constant. In such cases it is usual to calculate an average 'U value', to be applied to the whole roof.

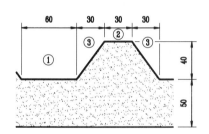

Insulation is Polyurethane.

In Zone 1, $U_1 = 0.413\text{W/sq.m°C}$
In Zone 2, $U_2 = 0.236\text{W/sq.m°C}$
In Zone 3, $U_3 = \dfrac{0.413 + 0.236}{2}$

$= 0.325\text{W/sq.m°C}$

With a temperature difference of 1°C, and for a 1m length of panel:–

Heat loss through Zone 1 = 0.413 x 0.06   =   0.025 W

Heat loss through Zone 2 = 0.236 x 0.03   =   0.007 W

Heat loss through Zone 3 = 0.325 x 0.03 x 2 =   0.019 W
                                                 ─────────
                                                  0.051 W

So heat loss through a 1m width, for a temperature drop of 1°C

$$= \dfrac{1000}{150} \text{ x } 0.051 \quad = \quad 0.34 \text{ W}$$

So effective U value is 0.34W/sq.m°C

Figure 8.5   The calculation of effective thermal transmittance for a non-uniform thickness of insulation.

An example of such a calculation is given in Figure 8.5. A composite panel has a profiled outer sheet, and a liner sheet which is virtually flat; it is completely filled with polyurethane insulation. U at the crowns is much smaller than U at the troughs, and varies continuously within the sloping rib sides. However, U at crown and trough is easily calculated according to the method already established, and within the rib wall zone the value will be taken as the average of the crown and trough figures.

The U value of a roof construction is the number of watts which escape through one square metre when there is a one-degree temperature differential. A square metre of the composite panel may be divided into an area of crown, an area of trough, and an area of rib wall. The heat loss through each can be calculated, and the three figures added; the result is the effective U value for the composite panel.

There are three basic choices for the position of the insulation within the roof construction, and these are shown in Figure 8.6. A *cold roof* has the main structural deck above the insulation. A *warm roof* has the main structural deck below the insulation, and an *inverted roof* has the weatherproof membrane below the insulation.

Cold roofs can include profiled metal with a cavity, or part cavity. They can also include a roofing system laid on joists, where the ceiling is fixed under the joists, and the insulation laid directly over the ceiling. Another very common cold roof arrangement is the typical pitched roof over a dwelling; the most common position for the insulation is directly over the ceiling.

In cold roof construction, the structural deck supports snow, wind and foot traffic, without these first being transmitted through other elements of construction. This makes for structural economy, and allows the use of very soft insulation, which is often the lowest in price. In the case of a pitched domestic roof, insulating at ceiling level uses less material than insulating in the slope; there is also economy in use as there is no requirement to heat the roof void (i.e. the loft space).

In the case of a cold roof, the inside surface can be selected on aesthetic grounds. However, a warm roof has its main structural support on the inside; if aesthetics have a high priority, the structural deck may require disguise.

The great advantage of the warm roof is that the structural deck is protected from extremes of temperature. In a British summer it is possible for the outside of a dark-coloured, insulated roof to reach temperatures as high as 80°C. In winter the temperature of this same roof could be as low as -20°C. The thermal expansion and contraction, resulting from such temperature changes can be injurious to some materials. Under the insulation, the temperature will be fairly stable throughout the year, and the damaging effects of thermal movement may be avoided.

The externally applied loads must be transmitted through the insulation to the deck. It is therefore not possible to use soft or easily compressible insulation. If a profiled structural deck is used, the insulation must have the strength to span between the crowns of the profile, otherwise the insulation could collapse under foot traffic. The weatherproof membrane and insulation require some form of anchorage if they are not to be blown away by the wind. One solution is to use stone chippings as ballast, but this adds to the dead weight which must be supported, and this can increase costs.

The inverted roof is attractive in concept. Both the structural deck and the weatherproof membrane are protected from wide variations in temperature. In fact they are protected from most adverse influences, and the membrane is sheltered from solar radiation, frost, mechanical damage, birds and rodents, etc. It is almost inevitable that ballast must be used, otherwise anchorages would penetrate the membrane; sometimes the ballast takes the form of paving slabs when the roof is intended for pedestrian access. This ballast may have unacceptable consequences for the structure, particularly on large clear-span buildings.

The insulation would need to be carefully selected to support the weight of the ballast and to withstand extremes of temperature, as well as being effective in both wet and dry conditions. The other great criticism of inverted roofs is that, should there be a leak, its source will be extremely difficult to find. Lifting large areas of ballast and insulation would be both time- consuming and expensive.

If any of these three variations on roof type had massive advantages over the other two, it would be universally adopted as *the method* of choice for all roofs. As all three are in common use, however, it is safe to assume that each has sufficient advantages to commend its use in some circumstances.

Any choices relating to the positioning of the insulation should involve a full consideration of the effects of condensation. This chapter must be studied in conjunction with Chapter 9.

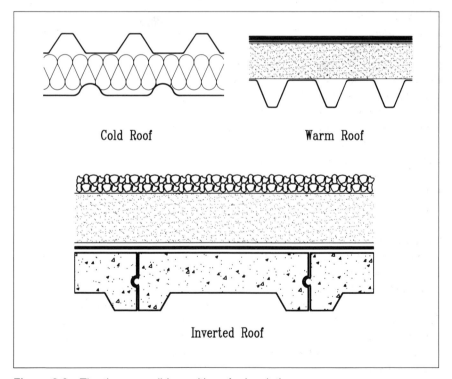

Cold Roof   Warm Roof

Inverted Roof

Figure 8.6   The three possible positions for insulation.

There is no single 'best' insulant, but it is possible to identify certain desirable properties. The most popular insulants may then be compared in these terms.

If a material is to be useful as thermal insulation, it must possess a low thermal conductivity. It will only be accepted if it can be purchased at a competitive price, and it should be light in weight in order not to overload the supporting structure. It must be relatively simple to install, sufficiently robust to withstand normal building-site usage, and able to survive damp or wet conditions. It must not be an undue hazard in the event of fire. The first six products in Table 8.1 all go some way towards meeting these requirements.

Phenolic foam boards or slabs are very effective as insulation, and can be of great assistance in keeping the thickness of construction to a minimum. Phenolic foam has a density of about 40 kg/m^3, and this means that a 50mm board weighs around 2 kg/m^2. The material is difficult to ignite and emits very little smoke or fumes. However, it is brittle and friable , and this can be a handicap in some applications. It is generally a little more expensive than other board insulants. It is available with a variety of facings including paper and metallic foil.

Polyurethane foam boards are only slightly inferior to phenolic foam in their insulating properties. They are of similar weight, and are available with similar facings. They have good structural strength, and this is sometimes enhanced by the inclusion of mineral fibres as reinforcement. However, their fire performance is unacceptable for some applications; they can be made to burn, and give off dense, poisonous smoke. Special treatments can bring about some improvement in fire properties.

Expanded polystyrene boards must be used in greater thicknesses to achieve as much insulation. They are light in weight, but have a fire performance similar to that of polyurethane.

All these boards have a closed-cell structure, and are not easily saturated. They can be degraded by ultraviolet light and should not be exposed to sunlight in the long term.

Mineral wool quilt is often about half the density of the foam boards but, as it must be used in double the thickness, its weight is fairly similar. It can be compressed to fit into awkward spaces, and is very easily cut or torn. It is not combustible, and cannot be induced to emit toxic fumes. Unlike boards, it cannot support itself between structural supportsand this continuous support must be given by some form of lining. In slab form it is stronger, but can become quite heavy.

Cork boards are an efficient, natural insulation; unfortunately they are   not available in sufficient quantities to meet all the needs of the modern building market. When properly cultivated, harvested and replanted, cork is an 'environmentally friendly' material.

Figure 8.7 illustrates some common applications for different forms of insulation. The examples are not exhaustive, and in some situations alternative insulations would be equally effective.

Polystyrene can be obtained in the form of large beads, about the size of broad beans. These are often used as a packing material, but are good insulation for filling difficult spaces. They are easily distributed between ceiling joists to fill the space

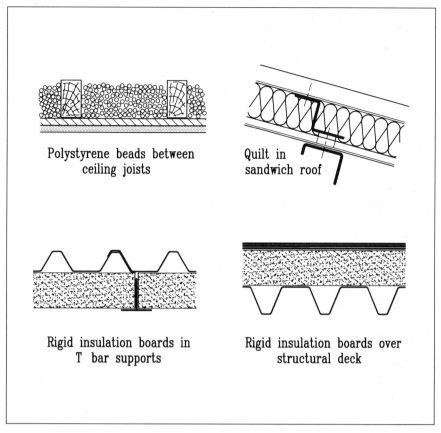

Polystyrene beads between
ceiling joists

Quilt in
sandwich roof

Rigid insulation boards in
T bar supports

Rigid insulation boards over
structural deck

Figure 8.7 Four arrangements for different types of insulation.

available. Vermiculite granules have been used in a similar manner, and mineral wool quilts are simple to fit between joists.

Industrial roofing often takes the form of a sandwich with quilt between two metal skins. The quilt is very easily compressed, so some form of rigid spacer is needed to provide a load path, and to ensure that the roof geometry is maintained.

Rigid insulation board, with a decorative underside, can be used without a lining sheet. The boards are supported in T-bars, and are strong enough to span between these supports. Special fixings are required for such applications, as the boards may compress very slightly, allowing the fixing screws to become loose.

Insulation boards, without decorative facings, may be laid over a structural deck to carry the weatherproof membrane. If the deck is profiled, the board does not receive continuous support; it should be strong enough to support foot traffic without collapsing into the troughs.

The Building Regulations permit the calculation of U values from standard data, and do not insist on testing to prove the result. They allow the use of light metal spacers, as shown in sandwich construction in Figure 8.7, but do not require their

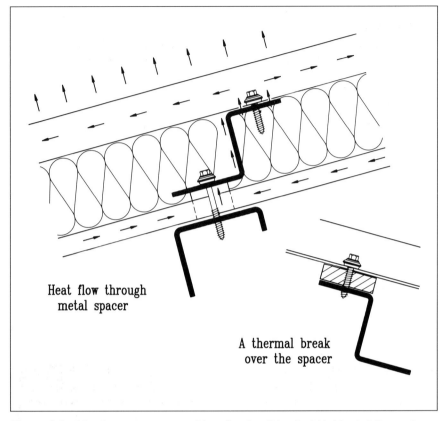

Figure 8.8   Metal spacers can provide a 'heat path' or 'cold bridge'. A thermal
                 break can reduce heat losses.

inclusion in the calculations. It is therefore often assumed that such 'cold bridges'
have little effect. In fact, the heat flow through such a spacer can be highly significant.
Drawing on established knowledge from a Swedish standard, it has been shown that
the use of thin steel spacers could increase the effective U value by as much as 50%.
A relatively thin thermal break could virtually eliminate the problem, however (see
D.T. Coates, 'How green was my spacer?', *Roofing, Cladding and Insulation*,
January 1990).

It is sometimes claimed that double-glazed rooflights have as good an insulating
effect as the area of roofing they replace. This is quite untrue! A double glazed
rooflight consists of an outer surface, an inner surface, and a cavity. The two skins
may be of thin plastic, or slightly thicker glass. In either case, the skins make little
contribution to heat retention. A calculation of U, ignoring the skins, would give a
value of 3.0 $W/m^2\,{}^{\circ}C$, and in practice the true figure is around 2.8.

The heat loss through normal rooflights is much greater than through modern
insulated roof construction. However, there are ways in which the U values of roof
lights can be improved. This will be discussed at greater length in Chapter 10.

STANDARDS
The Building Regulations Part L - Conservation of Fuel and Power.
BS 874:1973 - Methods for determining thermal insulating properties.

FURTHER READING
BRE Digest 108 - Standard *U* values.
BRE Digest 324 - Flat roof design: thermal insulation.

FLAT ROOF INSULATION

KORKTHERM
Natural Corkboard.
The economic answer
to roof insulation.
Proven track record
over last 50 years.

KORKTAPER FALL 1:40
Built in insulation
falls ensure a self
draining roof.   FALL 1:60

Agrement Certificate No. 86/1713.

KORKPLUS
Korktherm natural corkboard is inseparably
locked to Urethane foam to provide exceptional
roof insulation and maximum security against
wind uplift.

VENTSULATION
Korktherm natural
corkboard is
inseparably locked to
Urethane foam to provide
maximum security to wind
uplift and has vapour escape
channels and built in vapour
control layer to protect
against vapour
pressure/moisture.

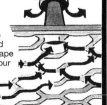

TOPFOAM
SUPERIOR
INSULATION
FOR THE
INVERTED
ROOF
Superior compressive
strength.
Minimal water
absorption.
CFC FREE.
Agrement Certificate No. 90/2486

Euroroof Ltd   Denton Drive, Northwich, Cheshire CW9 7LU
Tel: 0606 48222 Fax: 0606 49940

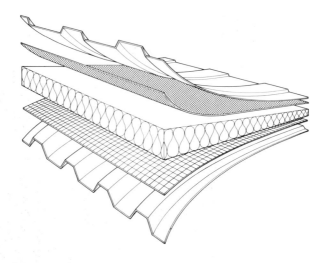

VAPOUR CONTROL

Plate 9.1   Laying a breather membrane over quilt insulation, in a double-skin profiled metal roof. (By courtesy of Alcan Building Products.)

Condensation is a very familiar phenomenon, recognizable as water droplets on cold surfaces. Typical examples include single-glazed windows in winter, and glasses holding iced drinks in summer; in either case the water droplets may be so small as to appear as a misting of the surface, or so large as to become streams of water running down the surface.

Another example is the ice which forms on the coil of a refrigerator or freezer. This is, in reality, a two-stage process; water droplets form on the cold coil, and these immediately freeze to form an ice layer.

In its effects on roofing, condensation has only recently been perceived to be a problem. The growing use of thermal insulation has resulted in condensation forming out of sight, within the roof construction. If this condensation is allowed to continue, undetected, it may lead to rust, rot, decay, mould growth, staining and loss of thermal insulating properties.

At first sight this is an illogical state of affairs. It may be argued that condensation occurs in large volumes on single glazing, but it is rare to see condensation on double glazing; the argument continues on the lines that the introduction of insulation has reduced the incidence of condensation - the opposite of what is claimed for roofing.

In fact this is not a sound argument. Condensation on either form of glazing occurs at the inner surface, but the damaging condensation in roofs occurs between the inner and outer surfaces.

Any proper appreciation of the incidence of condensation must depend on an understanding of the physics which govern this effect. This in turn requires a knowledge of certain properties and definitions.

Air is very seldom completely dry; it usually contains some water in the form of water vapour. There are various ways of expressing the degree of wetness of the air; one way is to state the number of grams of water in a litre of air (or in a cubic metre). However, the actual weight of water is not necessarily of great interest; it is often better to know whether the air is capable of holding any more water, and if so how much.

At any given temperature, there is a maximum quantity of water which can be contained in a given volume of air. If the air holds this maximum amount of water, it is said to be *saturated*. Air may be in any condition from completely dry to completely saturated, so its wetness may be described by means of *relative humidity*. The relative humidity is the ratio of water contained in the air, to the quantity which would be contained if the air were saturated; it is expressed as a percentage.

It was specified in the previous paragraph that saturation depends on a given temperature. Temperature has a significant effect on the amount of water which air can hold. If air is heated it becomes able to hold more water, i.e. it requires a greater quantity of water to become saturated. This is why most drying equipment is based on a blown stream of warm air; the warm air has a greater capacity for taking in water.

The relationship between temperature and relative humidity must be clearly understood. Relative humidity is the ratio of water held, to the maximum amount which could be held *at that temperature*. If the temperature of air is raised, the amount of water held is unchanged, but the amount required to saturate the air is increased. The change in the ratio of two quantities means that the *increase* in temperature causes a *decrease* in relative humidity.

Of course, a reverse effect is achieved if temperatures are reduced. The amount of water held remains constant, but the maximum possible amount reduces as the temperature falls, so the relative humidity increases. If the temperature continues to fall, the relative humidity will eventually rise to 100%. The air will then be holding the maximum possible amount at that temperature, i.e. it has become saturated. The temperature at which the air becomes saturated is known as the *dew point*.

If the temperature continues to fall, something different must happen — it is not possible for air to have a relative humidity of more than 100%. In fact, the relative humidity remains at 100%, but as the temperature falls, the amount of water to cause saturation becomes less. The surplus water cannot be held as vapour, it is expelled from the air as water droplets. This is condensation.

Condensation on windows can now be explained in terms of relative humidity and dew point. The warm air inside a room will hold a quantity of water vapour. If the outdoor temperature is low (e.g. freezing winter conditions), the glazing will be cold. The air in the room will circulate due to convection and draughts. As air passes close to the window, it is chilled; if its temperature falls below the dew point, some water droplets will be expelled, and these will be seen as moisture on the surface of the glass.

Double glazing provides a degree of insulation, so its inner surface will be warmer than that of single glazing. This explains why single glazing is more vulnerable to condensation.

Table 9.1 Typical temperatures and relative humidities for different types of occupancy.

Type of occupancy	Temperature °C	Relative humidity (%)
Dwelling — dormitory	15	50
Dwelling — washing/cooking	20	85
Office	20	40
School or college	20	50
Warehouse	15	30
Factory	15	35
Textile factory/paper mill	20	70
Swimming pool	25	70

Most air contains some water vapour, but human occupancy increases the initial content in numerous ways. People generate water vapour by respiration and through perspiration. A sleeping adult will produce about 40g of water vapour per hour, or about a third of a litre during eight hours' sleep; the source of condensation on a bedroom window is now obvious. The rate of moisture generation will increase by up to 50% as the activity level increases; it could reach 60g per hour for an adult playing badminton or squash.

Domestic activities, such as washing clothes, bathing, cooking, dishwashing and drying clothes, all create large amounts of water vapour. Similar actitivies also occur to a greater or lesser extent in offices, schools and factories, and some industrial processes, such as paper and textile production, generate massive amounts of vapour. The burning of fossil fuels also produces water vapour; oil and gas are the greatest contributors, but coal and coke are far from insignificant.

Table 9.1 summarizes operating conditions, for different types of occupancy, in terms of temperature and relative humidity. The figures can only be approximate, as no two dwellings, schools or factories will be identical, nor do conditions remain constant throughout the day or from one day to the next. However, the table gives a reasonable indication of the conditions which may prevail in different circumstances.

The condensation which occurs on windows is termed *surface condensation*; the same term would be used for condensation forming on a wall or ceiling. This is relatively rare, because walls and ceilings are usually insulated (the analogy to double glazing is again useful), but surface condensation is occasionally encountered in bathrooms. It is also possible to find local areas of surface condensation caused by 'cold bridging'; this could occur on a metal lining sheet, in a factory, where metal spacers bridge the insulation.

Not only is surface condensation relatively rare, it is also relatively innocuous; moisture on a bathroom ceiling soon disappears when the bath is emptied and the door or window opened. To promote rot or decay it would be necessary to minimize ventilation, and to make particularly heavy use of the bathroom.

There is, however, another form of condensation which is potentially far more damaging, namely *interstitial condensation*. As its name indicates, interstitial condensation forms within the construction, where it is neither visible nor accessible. This form of condensation can accumulate to cause long-term problems.

Interstitial condensation is not a simple topic, and various theories have been evolved to explain and predict its incidence. Knowledge and understanding of this type of condensation has grown as the use of thermal insulation has become more widespread; as explained in Chapter 8, this really means the last 20 years or so. It is perhaps best to start with the simplest explanation. This can then be developed and refined.

In Chapter 8 an example calculation was given for heat flow and temperature gradient through a multi-layer construction. The construction and calculated temperature gradient are shown in Figure 9.1; this time a further addition has been made. It has been arbitrarily assumed that the dew point is at -5°C, and the line representing this temperature has been drawn on the graph of temperature gradient.

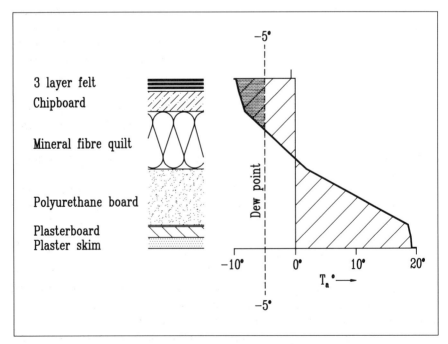

Figure 9.1 A simplistic view of the formation of interstitial condensation.

Of course, condensation at -5°C will occur as ice. When the temperature rises the ice will melt.

It has already been established that when air is cooled below its dew point condensation occurs. In Figure 9.1 the upper part of the roof is below the dew point. This includes the felt, the chipboard and the top few millimetres of the mineral wool quilt. Air in this region will give up moisture as condensation. The layers of felt are bonded together and there is no access for air, but the bottom layer will be laid on, nailed or partially bonded to the chipboard, so there will be a minute air gap between the board and the bottom layer of felt. Condensation can form in this air gap, and this will wet the chipboard.

The chipboard also holds some air within its construction, and there are air gaps at its joints. These are further positions where condensate may form. Mineral wool quilt contains a lot of air, so this provides further potential for the formation of condensate.

Until recently this was believed to be the only mechanism for interstitial condensation, but it is now appreciated that this is a gross over-simplification. The analysis must be extended to show how the moist air reaches the cool zones of the roof, and reaches these zones in sufficient quantity to deposit significant volumes of water. As the amount of air between the felt and the chipboard must be very small, the amount of water it can hold must be smaller still; there must be a dynamic effect which allows moist air to move to the cool zones, to deposit more condensate.

Experiments and measurement have shown that humid air exerts a greater pressure than dry air. This is most easily understood by first imagining a container full of dry air; now imagine adding some water vapour to the container, unless the container increases in size its contents must now be under greater pressure. It was stated earlier in this chapter that the occupants of buildings carry out a whole range of activities which add water vapour to the air; it follows that the air inside the building is likely to contain more water vapour than the air outside *and will therefore be at greater pressure.*

If there is a greater pressure inside the building, the internal air will attempt to escape through any holes, openings or gaps which are available. This will inevitably lead to warm, moist air entering the roof space.

There is another way to imagine what is happening, and this may assist in forming a clear picture of the process leading to condensation deposits.

One of the most popular experiments in junior school science consists of collapsing a can by atmospheric pressure. About 100ml of water is poured into a can of around 2.5 litre capacity and the cap is fitted with a piece of rubber tubing. The can is placed over a bunsen burner and heated until the water boils; at this stage there is a powerful jet of steam through the rubber tube, and it may be assumed that the can is full of water vapour. The burner is switched off and a clip fitted over the rubber tube to close it completely. As the can cools, the water vapour condenses, forming a partial vacuum within the can; the can collapses because the atmospheric pressure on the outside is not balanced by an equal pressure on the inside.

With this in mind, it is easy to imagine that when condensate is formed, in the cool zones of the roof, there is a local drop in pressure due to the removal of water vapour.

A minute distance away, the air is warmer, and as it still contains a large quantity of water vapour, it is therefore at higher pressure. The high-pressure air will migrate to where the pressure is lower, and more condensate is formed. The process is continuous as long as the temperatures and relative humidities are maintained.

Having established that condensation is caused by a temperature gradient, together with vapour pressure, it is possible to devise ways to combat its formation or its effects.

A form of construction which will not allow the passage of humid air is obviously desirable. A simple example of this is the fully filled composite panel; this panel has two metal skins and an infill of closed-cell foam insulation. The metal skins prevent air from reaching the foam, and the closed-cell foam structure prevents any movement of air within the foam. Interstitial condensation cannot occur in such panels (but may be able to occur at the joints). These panels cannot form a complete solution to all roofing applications, unfortunately they are relatively heavy to handle, are subject to thermal stresses, require through fixings, include compressible insulation, may have poor fire characteristics, and are available in a limited range of profiles and finishes.

Greater flexibility of choice is possible when a *vapour check* is used. A vapour check (sometimes called a *vapour barrier*) is a layer which is introduced into the construction to withstand the vapour pressure, and prevent humid air from gaining access to the cool zones of the roof.

Materials vary greatly in their ability to resist the passage of vapour. In order to compare vapour check materials, Table 9.2 lists some of the possibilities and gives their vapour resistance. The units adopted for vapour resistance MNs/g.

It must be clearly understood that the figures are applicable to a single piece of material in isolation. Thus, metal sheet is extremely resistant to the passage of vapour, but several metal sheets with unsealed overlap edges may allow vapour to pour through at the joints. In any vapour-resisting system, the quality of the joints is of vital importance. The reason for preferring the term 'vapour check' to 'vapour barrier' is now apparent. No material is a complete barrier against vapour, but all materials check it to some extent.

The example given earlier in this chapter predicted the occurrence of condensation purely on the basis of temperature gradient. A moment's thought will show that this is an incomplete approach. If a continuous metal foil were to be introduced on the warm side of the insulation, the vapour would be denied access to the cool zones of the roof, and condensation could not occur. The metal foil would have no discernible effect on the temperature gradient, however.

The concept of a continuous metal foil is largely theoretical. In practice, any vapour check has joints and weak points through which some vapour escapes. (This can be accommodated by a calculation method described in British Standard BS 5250. The method takes account of temperature gradient, vapour pressure and vapour resistance, but is beyond the scope of this book.)

Knowledge of the condensation process is growing rapidly. BS 5250 was published in 1989, but already there are improvements in the calculation methods. BRE Digest 369 provides an alternative approach.

Table 9.2 Vapour resistance for a range of roofing materials.

Material	Vapour resistance MNs/g
Aluminium foil	1000
Asphalt	10000
Chipboard in 20mm thickness	10
Closed cell foam in 50mm thickness	20
Concrete slab in 100mm thickness	20
Metal sheet	10000
Polyester film in 0.2mm thickness	250
Polythene in 250 gauge	500
Polythene in 500 gauge	1000
Roofing felt laid in bitumen	1000
Sarking felt	50
Woodwool slabs in 1oomm thickness	2

Vapour checks are located somewhere between the primary structure and the outer skin of the roof. Wind and snow loads, which are applied to the outer skin, must be transmitted to the primary structure. For this reason the vapour check is often penetrated by vast numbers of screws or nails, and each penetration increases the condensation risk.

Most of the materials listed in Table 9.2 may be used as vapour checks, but there is another possibility. Some manufacturers produce materials specifically for use as vapour checks. Such products can incorporate special features; they can be reinforced to improve tearing strength, made from specially formulated plastics to reduce the risk of pinholes, laminated from layers of plastic and foil, and produced in substantial widths to minimize site jointing.

Manufacturers of purpose-made systems usually offer a jointing system. These can utilize an adhesive to bond the overlapping edges, or the edges may be welded by heat, or by chemical reaction.

It is sometimes proposed that there is no need to use a separate vapour check with a profiled metal lining sheet, as the laps can be sealed to create a vapour check. This is a reasonable argument, but it can prove difficult to make good seals in the metal lining — and it is much more difficult to test the seals!

It is sometimes possible to seal around fasteners by means of adhesive. Alternatively, it is now possible to buy fasteners with a coating of special sealant; such fasteners seal to the vapour check as they are inserted. The designer should always seek ways of minimizing the extent to which the vapour check is penetrated.

Despite the best efforts of the designer and roofing contractor, it is all too common for the vapour check to be damaged and reduced in efficiency. This could be the result of tearing, badly sealed laps, local flaws, or screw penetrations. As additional protection against such eventualities, it has become common to specify the use of *breather membranes* (sometimes called *breather papers*).

A breather membrane is most useful where there is an air space between it and the outer skin. For example, a breather membrane could be below the battens in a tiled or slated roof; it could also be immediately below a profiled sheet roof (the air space would occur only at the ribs).

A breather membrane is formed from a material which will not absorb moisture (e.g. polythene, foil, bitumen felt), but is perforated by vast numbers of minute pores. The pores may be formed by deliberate perforation during manufacture (as would be necessary for metal foil), or they could result from a weaving process (such as weaving plastic threads into a fine cloth).

The function of the breather membrane is very simple. If vapour has succeeded in penetrating the vapour check it will continue towards the cold outer skin, under the action of vapour pressure. The pores in the breather membrane allow the vapour to pass through virtually unimpeded. By comparison with the materials listed in Table 9.2, breather membranes should have a vapour resistance of less than 1.

When the vapour reaches the cold outer skin, it is possible that condensate may form. If it forms in relatively large quantities it may trickle down on to the breather membrane. However, the pores in the breather are so small that water cannot flow through them (this is a function of surface tension). The condensate is unable to wet the insulation; it may drain to the eaves where it can be discharged into the gutter, or

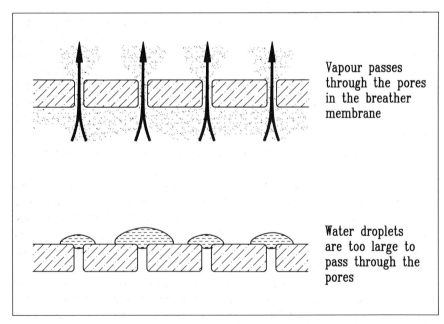

Vapour passes through the pores in the breather membrane

Water droplets are too large to pass through the pores

Figure 9.2   A breather membrane allows the passage of vapour, but prevents the passage of water.

can simply remain on top of the breather until a change in operating conditions allows it to be evaporated and ventilated to atmosphere. Of course, a rooflight in the middle of the roof slope would disrupt the drainage runs as the breather could not continue across a light.

Figure 9.2 shows how the breather membrane creates a one-way system for water vapour. In placing breather membranes, there is no requirement for elaborate jointing methods; simple overlaps are adequate. The laps should be arranged to shed water, not to allow it back under the breather.

The most positive way to reduce condensation risks is by ventilation. Ventilation can assist in several ways. In a warm room with high humidity (such as a bathroom or laundry), the water vapour in the air increases the vapour pressure. Simply opening a window, to increase ventilation, allows some of the humid air to be replaced with less humid air from outside; this reduces the vapour pressure. Unfortunately this also causes large heat losses. Some buildings use a sophisticated ventilation system in which heat exchangers are used to retrieve the heat from the air before it is removed from the building.

Ventilation is also used within the roof construction. In domestic applications it is usual to lay the thermal insulation directly over the bedroom ceilings; this means that the roof void is cold. Warm humid air will escape into the roof void, and this can deposit moisture onto a cool surface. By ventilating the roof void it is possible to keep the air in the roof void at a temperature reasonably similar to the external air. This greatly reduces the risk of condensation. The Building Regulations require a minimum level of ventilation to the roof void.

Profiled sheets are often laid over level insulation, i.e. the ribs are empty. Any vapour, finding its way through the roof construction, will arrive at the ribs and deposit condensate on the inside of the profiled sheet. Sometimes this will freeze and accumulate over a period of hours or even days. Eventually conditions change, the ice melts and the condensate runs into the insulation (or drains down the breather membrane). The situation can be much improved by the use of ventilated profile fillers at the ridge and eaves. Figure 9.3 shows some of the possible forms for ventilated filler pieces; in selecting a particular shape, consideration must be given to the location of the filler and the risk that the ventilation opening may become an entry point for storm water.

The notches in the fillers are very small compared with the area of some industrial roofs, but experience has shown that ventilated fillers make a significant contribution to reducing condensation. This is partly because wind creates high pressures and suctions around the eaves and ridge, and these pressure variations help to move air through the rib voids. Also the use of the notched fillers reduces the vapour resistance of the roof sheets, and the vapour escapes to the atmosphere.

It is advisable to use a 'notched base' filler at the eaves as this allows accumulations of condensate to drain into the gutter. Notched fillers can be produced with a fine mesh or gauze over the ventilation openings. These are useful for keeping out insects.

In membrane roofs, vapour pressure is sometimes relieved by means of small

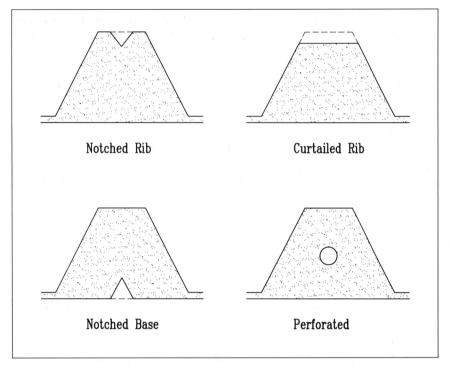

Figure 9.3   Profiled fillers can be modified to allow trickle ventilation.

purpose-made ventilators which are fitted at intervals. These are particularly beneficial in conjunction with insulation which allows the passage of air. One manufacturer
produces an insulation board with a grooved surface for just such applications.

Another approach to the condensation problem is the use of absorbent coatings.
The coating is applied to the back of a metal sheet in a thickness of about 1mm. It is
highly absorbent so, when condensation occurs, the condensate soaks into the coating
and is held there. When conditions change, the condensate evaporates and returns to
the atmosphere. These coatings are very useful for controlling occasional small
amounts of condensation; they are not capable of dealing with large volumes of
condensate (e.g. on the ceiling of a bottle-washing room in a dairy), or a continuing
build-up (e.g. when the cold outdoor conditions prevail for several days or weeks).

These coatings should be used with caution. If they perform their design function,
they will hold moisture against the metal for prolonged periods. This is rather like
applying a wet poultice; some metals will be corroded unless properly protected. Had
the condensate formed and dripped off, the metal would have been less at risk
(although the condensate may have posed a threat to another part of the construction).

The accumulation of condensate has been mentioned several times in this chapter,
and this is an important topic in its own right. When a particularly cold night causes
condensation for a few hours, this may be little more than a nuisance. The return of
normal conditions will soon dry up the moisture, and probably no harm will have

been done. However, if the condensate continues to form over a long period, it will saturate the insulation, timber structure, ceilings, etc. It may take several months of summer weather to dry out the accumulated condensation of the winter.

An even worse scenario is that the condensation in winter is so great that it has not all been dried by the end of the summer. This can only result in an annual deterioration leading to premature failures through rot and decay. BS 5250 provides a means of calculating the quantities of condensate formed, and assessing the possibility of harmful accumulations.

The design and construction of roofs must take account of condensation risks. New materials and new knowledge will certainly affect our view of this subject over the next few years; new ideas will appear in the technical press. In the meantime, the worst risks will be avoided by thoughtful design, accurate workmanship, and a little common sense.

STANDARDS
The Building Regulations Part F2 - Condensation.
BS 5250:1989 - Code of practice for control of condensation in buildings.

FURTHER READING
Thermal insulation: avoiding risks - BRE - 1989
BRE Digest 336 - Swimming pool roofs: minimising the risk of condensation using warm deck roofing.
Defect Action Sheet 59 - Felted cold-deck flat roofs: remedying condensation by converting to warm-deck - BRE - 1984

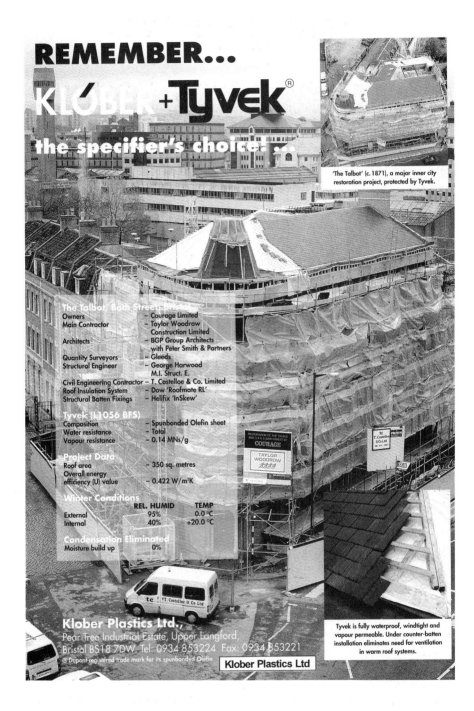

NATURAL LIGHTING

10

Plate 10.1   A distinctive pattern of glazed lights to provide a high level of
illumination to college classrooms. (By courtesy of David Duckham RIBA.)

Translucent materials are commonly incorporated into roofs in order to provide
lighting to the building. There are several popular materials, and their application
varies according to the nature of the roof.

Some systems depend on normal glazed windows, in the form of dormers or
monitors. These are good and successful methods of providing natural light, but do
not require explanation here; windows are well documented in other books on
building.

This chapter will concentrate on the type of daylight units which are purpose-made
for use in roofs; they usually follow the line of the roof (as opposed to the roof shape
being modified to accommodate vertical windows). These units are mainly used in
factories, offices, sports halls, shopping malls and warehouses. They do not feature
largely in domestic applications, although they may be of use in carports and
conservatories.

Four materials which are regularly used for daylight units: glass, PVC, GRP and
polycarbonate. Each of these materials has certain strengths and weaknesses which
are likely to affect its use in specific situations. There is no' best' or 'worst' material,
but there may be a particularly suitable material for a given detail.

Table 10.1 provides a simple comparison of the four materials. The numbers can
only be approximate as there are various formulae for plastic, different filters for
glass, a range of shapes to influence strength, and unique design loading require-

Table 10.1   A comparison of properties between the four most popular
rooflight materials.

	Glass	PVC	GRP	Polycarbonate
Light transmission	87%	82%	85%	90%
Normal thickness, mm	6.0	1.5	1.0	1.5
Min. safe temperature, °C	*	-10	-30	-40
Max. safe temperature, °C	*	+45	+120	+120
Service life	Very long	Medium	Long	Very long

** Glass can be used safely over an enormous range of temperatures, but is
vulnerable to sudden changes in temperature and to stresses induced by
restriction of thermal movement.*

ments for each building. It is also significant that the properties of plastics change
as they age; in particular they become less transparent, but they can also lose strength
or become brittle.

The service life indicated is simply a general statement. PVC loses its ability to
transmit light, and hence its value as a rooflight becomes questionable after 10 to 15
years, although it will function as weather protection for perhaps 20 years. GRP
suffers similarly, but retains its light-transmission properties for a little longer.
Polycarbonate retains its properties even longer, and has higher strength. Glass does
not deteriorate to any significant extent, but is more likely than plastic to be damaged
by impact or thermal shock. More will be said about these materials later in this
chapter.

Rooflights are widespread. This demonstrates that they are attractive to the owners
and users of buildings. It is worth examining the reasons why they should be so
popular.

The most obvious reason for installing rooflights is to provide lighting to the
interior of the building. This will not reduce the number of artificial light sources
required, but will reduce the number of hours for which they will be used. The next
most important reason is what is commonly called the 'feel-good factor'. It has been
shown that workers are more contented when they are in touch with their environment; they want to be aware that it is light, dark, sunny or dull. When people are cut
off from any contact with the outside world, they can become confused or unhappy.

Natural lighting can supplement artificial lighting by softening the shadows, creating a less harsh effect. Artificial lighting may be preferred where accurate work must be carried out, but natural light may be used in some other areas. Natural light can offer certain safety advantages in that it cannot be cut off by a sudden loss of power.

Quite aside from the ultimate use of the building, rooflights can confer advantages during construction. When a large factory is being built, it is in the interest of the developer to complete the roof and walls as early as possible in the construction programme. It is then possible for work to proceed inside the building, regardless of the weather conditions. However, the power supply is not usually connected until the building is complete, so there can be no artificial lighting unless the contractor makes special arrangements and provides generators. Rooflights can provide the necessary lighting.

Many industrial units are built by local authorities, government agencies or speculators. Tenants are found after the buildings are completed. It is usually left to the tenant to connect the power supply, and to install artificial lighting appropriate to his needs. As a result, the factory requires some natural light in order that prospective tenants may actually view the building. Again, rooflights are the source of this lighting.

The behaviour of rooflights in fires varies according to the rooflight material. This will be discussed in Chapter 15, but it should be noted that some rooflights have the capacity to disintegrate under the effect of fire, thus allowing the escape of smoke and fumes; such rooflights can make an important contribution to the safety of the building's occupants in the event of fire. Of the materials in Table 10.1, PVC and polycarbonate provide this self-venting performance.

It should not be thought that rooflights are a source of numerous benefits without cost or penalty. In fact there are many weaknesses inherent in the use of rooflights, but the advantages are usually held to outweigh the disadvantages.

The greatest condemnation of rooflights is in respect of heat loss. Chapter 8 included comments on the insulation value of roofing construction, and of rooflights. The current requirement, under the Building Regulations, is that the roofing must have a U value not exceeding 0.45 W/m^2°C, (or 0.25 in the case of domestic roofing). This requirement is waived for rooflights up to a stipulated area. However, for double-skin rooflights U is approximately 2.8, and for single-skin lights U is approximately 5.7. The different materials have little effect on the figure because the skins are relatively thin.

Remembering that U is a measure of the rate at which heat can escape, it is clear that heat can be lost much more rapidly through rooflights than through insulated roof construction. This is illustrated in Figure 10.1, which shows how the average value for U of the whole roof increases with the addition of rooflights. This produces the dramatic conclusion that 10% of single skin rooflights, or 20% of double skin, will allow heat losses through the roof to increase by 100%

Because the lights have poor insulation qualities they are likely to suffer from condensation on cold mornings. If there is enough condensate to drip into the building, or to wet the purlins, this can be a considerable nuisance.

Just as heat losses can reduce comfort levels in winter, so solar gain can do so in

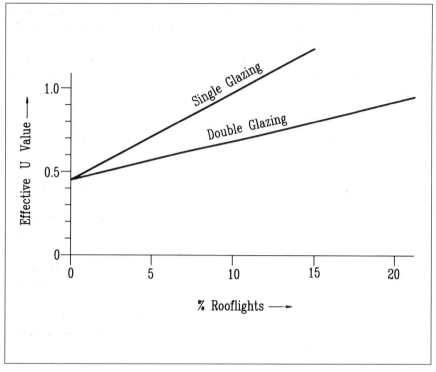

Figure 10.1   A graph to show how the percentage of rooflights can increase the
average U value of the roof (starting from U = W/m²°C).

summer. Sunlight includes radiant heat, which is short-wavelength radiation; this
radiation passes through rooflights and windows, and raises the temperature of
anything upon which it falls. When the contents of the building become warmer they
too radiate heat, but this radiation has a longwave length and cannot escape through
the windows and rooflights. This greenhouse effect causes the building to become
increasingly hot, and can lead to intolerable conditions.

GRP lights can be tinted to reduce solar gain and to soften shadows. This also
reduces the amount of light transmitted, but the gain in user comfort may be deemed
adequate return for the loss of light.

It should be remembered that the use of insulation in factory roofs is a relatively
recent development. When single-skin roofing was commonplace, there was no
need to consider heat losses through rooflights. Lighting was provided by replacing
some profiled sheets with translucent sheets, or by incorporating large areas of
glazing; the roof had a U value of about 5.7 W/m²°C whether the sheets were of glass,
plastic, asbestos cement or corrugated iron.

The introduction of insulation, as described in Chapter 8, and vapour control
layers, as described in Chapter 9, has dramatically changed this situation. Not only
does the area of the lights contribute to heat losses, as demonstrated in Figure 10.1,
but the presence of a rooflight means that the vapour control layers must be suitably

terminated and sealed. It should be clear that a breather membrane needs to be continuous from ridge to eaves; a chequer pattern of rooflights is not compatible with the use of breather membranes.

Sometimes a sandwich construction roof has the laps in the liner sheet sealed in an attempt to make the liner act as a vapour barrier. If the roof incorporates double-skin rooflights, translucent liner sheets may be used as the inner skin. The laps in the translucent liners must also be sealed and stitched to maintain the vapour check. The success of such an arrangement is largely dependent on the skill of the fixer, and on the design of sensible details. Lapping the plastic sheet over the metal, at sidelaps, is one way of improving the overlap quality.

Glass and plastic are not usually as strong or as stiff as metal, fibre cement or timber. It is possible that including rooflights may result in the need to reduce the spacing of the purlins.

There is also a safety consideration. Most roofing materials will support foot traffic without damage (or perhaps with some minor denting in the case of thin profiled metal). However, most rooflight materials will not carry the weight of a man; should a person accidentally step on such a rooflight, he may well fall straight through. This can be a very important consideration when people other than builders or roofers may have access to the roof.

When fixing plastic rooflights, it is usually necessary to use special large washers to prevent the lights being pulled off in high winds. This places extra responsibilities on the designer, buyer, fixer and supervisor.

There is also the question of durability. Elsewhere in this book it is shown that roofs in fully supported non-ferrous metals can last 100 years. A similar lifespan is possible for tiles and slates, while profiled aluminium and fibre cement roofs have achieved lifespans of over 50 years. Rooflights in such roofs should offer similar durability — or the designer should ensure that the building owner is aware that the rooflights may need to be replaced after a stipulated period.

Deterioration in a rooflight could mean that it became unable to keep out the weather, but another form of deterioration is a growing inability to transmit light. Figure 10.2 shows how the four most popular translucent materials deteriorate in this respect. Glass shows no significant change, and polycarbonate loses its light transmission qualities at a very slow rate. PVC is degraded by ultraviolet radiation, and its light transmission falls rapidly at first; GRP lies about halfway between PVC and polycarbonate.

GRP lights are formed from glass fibres in a matrix of resin. As they age their surfaces deteriorate and allow some fibres to project, these may become dirt traps. This can be prevented by maintenance coating to restore the intact surface. This coating can greatly extend the life of the light; it is usually required after 10–20 years, according to local conditions. Some modern GRP laminates with outer gel-coat surfaces will display enhanced durability.

The values indicated are approximate only. They depend on the exact composition of the material, and on the local conditions. A south-facing roof slope will receive more radiation than a north-facing slope, and some districts receive more sunlight, etc. Local pollution may be aggressive to some materials. Dirt deposits may reduce

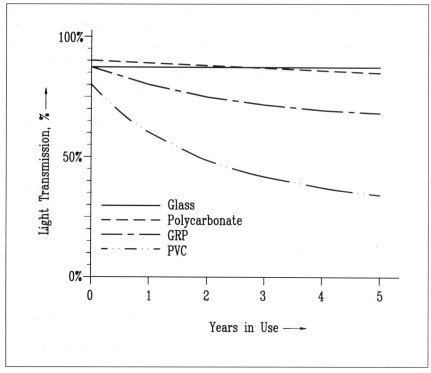

Figure 10.2   A graph to show how ageing can affect the light transmission of the four popular rooflight materials.

light transmission, but may also protect the rooflight from ultraviolet radiation.

Most modern rooflights are of double-skin construction; the values quoted in Figure 10.2 are for single skins. The inner skin receives some protection from the outer skin and does not deteriorate so rapidly. In designing a rooflight system it is important to decide on priorities. GRP is the most popular choice for its general all-round qualities, competitive pricing, and tolerance of mishandling. If illumination during construction is the main requirement, then PVC lights could meet this objective. If long-term illumination is needed, then glass or polycarbonate offer the best solution.

It is not easy to define what level of natural lighting is required, but one method uses the concept of *Daylight Factor*. This is defined as the percentage of light inside a building compared with that outside. It is claimed that a daylight factor around 6% is required for factories, and 3% for warehouses. The light transmission through double-skin lights cannot be better than 81% (i.e. 90% x 90%), so a 6% daylight factor will require at least 7.5% rooflights. This can lead to significant heat losses, as shown in Figure 10.1

It is possible to use triple-skin rooflights, or to incorporate insulation into double-skin lights. This improves thermal insulation, but at the expense of light transmission, so the benefits are limited. As awareness grows that energy losses are high through rooflights, the current trend is towards smaller areas of lights. In fact, 5% rooflight area is becoming fairly common, and this may be in special insulated lights.

It is possible to construct triple-glazed glass lights, or double-glazed polycarbonate with translucent infill, with a U value of under 2 W/m^2°C, while maintaining a light transmission of 60% or more, and 5% of these rooflights would provide all the advantages of free daylight: worker contentment, background lighting, construction lighting, tenant viewing and safety in a power cut. The energy losses are quite small for such an arrangement, but the saving in heating costs must be offset against higher initial cost.

The other perceived disadvantages of rooflights may be addressed by reference to the available materials.

Smaller areas of rooflights will automatically reduce the potential for solar gain. If large areas of rooflights are essential, then these are best arranged to face north — hence the familiar north-light roofs. There is a possible case to be made for east-facing rooflights; these will allow early morning sunshine to heat the building, but will not continue to increase the temperature as the sun moves further south.

Disruption of vapour control layers can be reduced by using rooflights in narrow bands starting from the ridge. They should not continue all the way to the eaves, as this would introduce risks for personnel walking on the roof (it would encourage people to step on to rooflights). An even better solution is to use a continuous rooflight at the ridge, and no rooflights in the slopes. This arrangement also allows the strength of the roof sheets to be fully utilized, as spans are not restricted to suit weaker plastic materials.

Safety can be improved by the use of polycarbonate lights, as polycarbonate is stronger than the other materials. A further safety measure can be to place the rooflights on kerbs, which makes them much more visible and less likely to be stepped on by accident. Lights on kerbs are also generally more easy to seal against the weather.

The various forms of rooflights each have advantages and limitations, and they will be discussed in general terms.

The traditional rooflight system for industrial roofing was 'patent glazing', and this system is still in widespread use. The system depends on glazing bars spanning between purlins, supporting glass or plastic windows. The original glazing bars were made of iron or steel, but modern ones are usually extruded aluminium. Figure 10.3 shows typical glazing bars for both single and double glazing; these bars must be fixed to the purlins by means of special attachments which allow free thermal movement. The glass is sealed by rubber gaskets or by a similar method, but the bars include drainage channels as a second line of defence. It is acknowledged that the gaskets could never provide 100% weather protection.

In fact, when the concept was first developed, inventors took out numerous patents for various shapes of glazing bars. It proved very difficult to protect an idea fully, and new patents were taken out to defeat previous ones. The result was that the different manufacturers were soon forgotten, and the systems became known collectively as 'patent glazing'.

Many early commercial buildings were constructed with north-light roofs, these

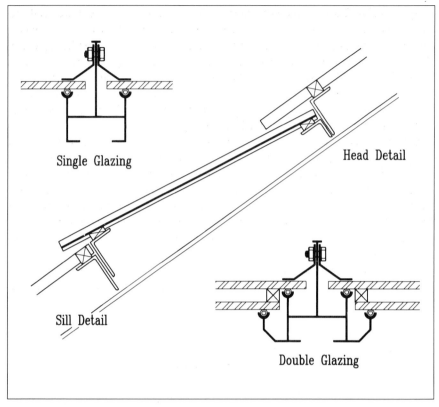

Single Glazing

Head Detail

Sill Detail

Double Glazing

Figure 10.3 Typical patent glazing details.

had patent glazing to each entire north-facing slope. This gave rooflight areas of
perhaps 20%–30% of the roof area; the result was extremely light working conditions. The roofs had no thermal insulation, however and the buildings were difficult
to heat. When such buildings are refurbished, or if new ones are designed, a different
approach is necessary. The areas of patent glazing are reduced by including runs of
insulated roofing between patches of rooflights, and the new lights are double-
glazed.

When patent glazing is incorporated into a roof slope, it must be so arranged that
the upslope roofing drains on to the glazing while the glazing drains on to the
downslope roofing. Typical details are shown in Figure 10.3. The pitch of the glazing
is significantly less than that of the remainder of the roof, and this imposes limits on
the applications which are possible.

While it is possible to use plastic in patent glazing, glass is more usual. When patent
glazing was at the height of its popularity, the glass was almost always *Georgian
wired*, incorporating a reinforcing mesh of fine wires; the reinforcement helped to
prevent the glass from shattering in fires, and it also reduced the risk of broken glass
cascading into the factory after accidental damage.

Georgian wired glass is not a decorative product, though it is quite suitable for factory roofing. Today its market share is in decline; when aesthetics are considered there is a clear preference for laminated or toughened glass. Both of these products meet the safety requirements, without the obscuring effect of the reinforcing mesh.

The most attractive features of patent glazing, to a roof designer, are its durability, its consistency and its adaptability. Glass and extruded aluminium are both capable of surviving for more than 50 years with very little maintenance. Glass does not show any significant change of properties over decades of use, and tapered glass panels allow complex shaped areas of rooflights to be produced.

Factors which have led to its loss of market share are cost, compatibility and complexity. Patent glazing is likely to prove more expensive, initially, than most modern plastic products, and its extreme durability is of no great consequence to a designer planning a factory for an intended 20-year life. The head and sill details shown in Figure 10.3 would not fit comfortably into a modern insulated roof with a breather membrane. Patent glazing usually requires additional items of structure and special flashing details at junctions; some modern plastic products are much simpler in application.

GRP profiled rooflights are available in practically every roof sheet profile. They are regularly used with steel, aluminium, fibre cement and bitumen-based profiles.

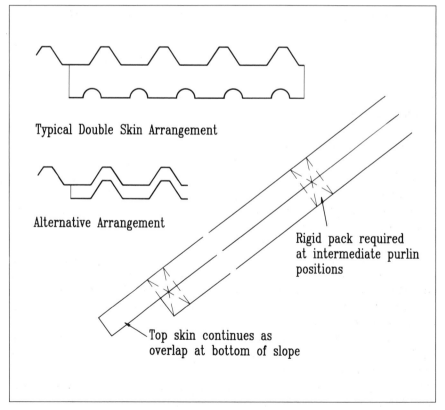

Figure 10.4   Typical details for double-skin profiled rooflights.

They were originally developed for use with asbestos cement sheets, and were laid in patches three sheets wide by one purlin spacing deep giving an overall light size of approximately 5m².

The modern trend is to use a single 1m width of translucent sheet 5m or 6m long. These still conform to the 5m² area, and are more easily accommodated in insulated roofs with vapour control membranes.

The other modern development is the use of double-skin rooflights. Figure 10.4 shows two possibilities. The lights may be factory-assembled, and sealed, or site-assembled. Good factory-assembled lights are probably the best option as they are securely sealed against moisture (which would lead to unsightly condensation inside the lights). Site-assembled lights cannot be completely sealed and will almost certainly suffer occasional condensation; however, they also allow vapour to escape easily, whereas poorly sealed factory-assembled lights tend to trap moisture, causing prolonged condensation problems.

Double-skin lights can also be created on site by simply fitting translucent liner sheets to coincide with translucent top sheets. Care is needed in setting out, as 'creep' could result in the two layers becoming out of sequence. Special attention is required around the edges to prevent insulation quilt spreading into the light area, and all laps should be sealed and stitched to reduce vapour transmission.

The sheets can be lapped at sides and ends, to each other and to the profiled roof sheets. It must be appreciated that the rooflights and profiled sheets are made from different materials, in different thicknesses, on different machines and by different processes, and often in different factories. It is unavoidable that tolerances will lead to a small degree of lack of fit; laps will usually require sealant.

At the purlin positions, stiff packs are necessary to prevent the fixings from crushing the lights. This information will be required by the manufacturer when ordering factory-assembled lights. All fixings will require large washers as protection against 'pull through' under wind load.

GRP lights are economical, readily available and easy to install. They have the durability to provide a 20-year rooflight, although they may become dirty and discoloured with poor light transmission as they age. Their properties in fire are dependent on the resins used in their manufacture, and most manufacturers provide comprehensive product information on request.

It is perhaps unfortunate that downward cost pressures from developers have led to GRP lights being produced in typical thicknesses of around 1mm. A 25% increase in thickness, together with a modern gel-coat finish, would not have a major effect on rooflight costs, but would significantly improve both the strength and durability of the product.

GRP lights are formed in a continuous process. This is very economical for straight sheets, which are simply cut to length at the end of the production line. Lights which have more complicated shapes can be hand-made in GRP, but this is time-consuming and expensive.

Figure 10.5 shows a cross-section through a rooflight designed to be mounted on kerbs. The light may be a square or circular dome, or it may be continuous as a barrel vault ridge (in which case there may be minor profiling across the light to increase strength and simplify jointing). Forming the kerbs is an extra complication in the roof construction, but it makes the light independent of the roof sheet profile, simplifies the seals around the light, and reduces the risk of anyone accidentally stepping on the light.

These lights cannot be formed in a continuous process; the most common method of manufacture consists of forming a plastic sheet around a mould under a combination of heat and pressure. PVC and polycarbonate lights are both formed in this way, usually in fairly small modules to keep down the costs of the moulds and the plant.

PVC gained popularity because it is self-venting in fires. In fact, it is easily damaged by heat, and PVC rooflights should not be fixed into dark-coloured metal roofing because these can attain temperatures above the safe working temperature for the plastic.

Polycarbonate is steadily increasing its market share as its durability, strength and stability become more widely appreciated. Its initial cost is likely to be higher, but this may be offset against its considerably greater service life, and its ability to transmit more light (possibly allowing the use of smaller areas). It is self-venting, but withstands greater temperatures than PVC.

Polycarbonate is usually specified for applications where there is a perceived risk of vandalism, and it is similarly used in high-security situations such as prisons or some military establishments.

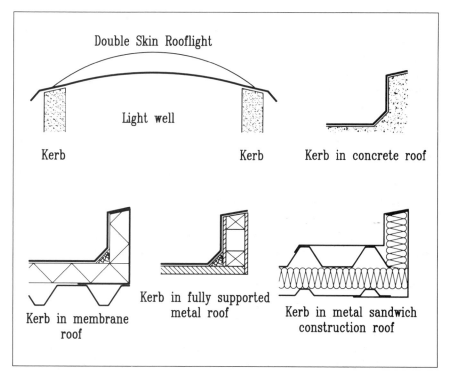

Figure 10.5 Various forms of construction for rooflight kerbs.

Rooflights can confer many benefits, but they can also introduce numerous disadvantages. An appreciation of the principles involved, and the materials available, allows the designer to maximise the benefits while minimizing the disadvantages.

It is to be regretted that there are some roofs in which the designer has used rooflights as a central aesthetic feature, without regard to the true needs of the building's occupants. It is certainly possible to make attractive use of rooflights, but the genuine lighting needs should always be the starting point in the design process.

STANDARDS

BS CP 153:Part 1:1969 - Windows and rooflights - cleaning and safety.
BS 4154:Parts 1 & 2:1985 - Corrugated plastics translucent sheets made from thermo-setting polyester resin (glass fibre reinforced).
BS 5516:1977 - Code of practice for patent glazing.
[N.B. the underside of a rooflight is also a 'lining', and is subject to the same regulations and standards as roof linings - see Chapter 7]

FURTHER READING

Setting the Standard - Patent glazing:Notes for the guidance of specifiers - The Patent Glazing Contractors Association.
Special Application Guides - The Glass and Glazing Federation.
Trade literature.

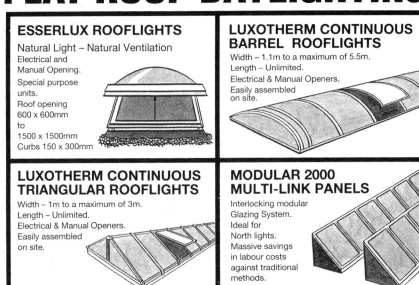

ROOFLIGHTS & TRANSLUCENT SHEETING

DAYLIGHT is a key element in the design and function of all buildings. GRP (glass reinforced polyester) sheeting has for many years been the preferred choice for admitting daylight to industrial and commercial buildings- either as wall or rooflights. **LITAGLASS** sheeting has been made for this purpose by **HARTINGTON CONWAY** for over 40 years, and is still the leading choice. Other plastics come and go, but LITAGLASS GRP sheet continues to have the edge- it is

TOUGH   EASY TO FIX

RESISTS HEAT, WEAR ,CORROSION

LITAGLASS GRP sheet can last as long as most cladding systems, when properly specified and fixed. As with most products, short-cuts in fixing and selection of thin and cheaper sheeting can also lead to shorter life and earlier replacement.

LITAGLASS sheet comes in
> 3 Fire resistant grades
> 3 Weights
> 3 Surface finishes

Fire Grade   This usually depends on location and regulations. Class3 SAB sheet is adequate for most external cladding, is cheaper, and is more resistant to U/V light. Class 1 and Class 0 fire retarding sheet is more expensive, and is usually required for internal linings.

Weight   Standard weight GRP sheet is 6oz/squ.ft or 1.83Kg/s.M. 5oz sheet is available, but is not recommended for anything but agricultural use. A heavy-weight 8oz sheet is also available for a small extra cost, and gives much improved appearance, strength, life and handling.

Surface protection Polyester 'Melinex' film has become standard surface protection in the UK over the last eight years. However, it has a limited life in strong sunlight. Gel-coat is used in sunnier climates, and is now available from Hartington Conway at similar cost. For almost indefinite translucence, specify LITAGLASS 'Diamond' Grade, with fluoride based protective film.

FIXING
Good fixing and handling will lead to a long life and trouble-free roof. Key points should be specified, such as sealing of all liner panel side and end laps, correct stitching of side-laps, correct main fixings at purlins, and correct spans.

HARTINGTON CONWAY offers the best SERVICE in the industry for all Rooflight and translucent sheeting needs, with daily deliveries nation-wide from our Coventry premises. All sheeting is made to order. Phone for technical help or literature.

HARTINGTON CONWAY LTD,
Aldermans Green Industrial Estate Coventry CV2 2QU
Tel 0203 602022   Fax 0203 602745

ROOF OPENINGS

11

Plate 11.1 A site-welded soaker was necessary where this pipe penetrated a concealed fix roof. (By courtesy of Melvyn Rowberry Welding Services Ltd.)

The term 'roof opening' is not intended to suggest a hole in the roof; rather, it should be taken as describing a disruption in the normal roof surface by some form of penetration. For example, chimneys, vent pipes, rooflights, access hatches and ventilators, all necessitate roof openings.

Most roofing materials and roofing systems are designed and developed for application to unobstructed areas. Tiles, membranes and profiled sheets can be applied to any plain, continuous surface; each cladding unit follows the previous one, and there is no physical limit on the area to be covered. However, the introduction of openings can be a severe complication, and the roofing system cannot be successful unless it offers solutions to the problems caused by openings.

The complications created by openings can arise from a variety of sources, and these will be discussed before considering standard details for specific roofing materials.

The most important property of a roof opening is its size. A very small opening to accommodate a vent pipe is usually simple to design and construct, but a large opening for a glazed light will require structural support, extensive weathered joints, alternative drainage runs and will need two different tradesmen working on the same roof.

Roof pitch is also of great importance. The rate of drainage is far higher when the pitch is steep, so it is easier to make weathertight details in steeply pitched roofs. Openings in tiled roofs are relatively simple to achieve, because the pitch is likely to be over 20°. An opening in a secret-fix roof, at 1.5° pitch, is more difficult to weather.

Profiled sheets and profiled tiles effectively divide the roof surface into a series of drainage channels separated by ribs or crowns. In a normal application a channel runs from ridge to eaves, but at an opening some channels end. A means must be found to direct the water in these channels around the opening, and on towards the gutter.

It follows that the position of an opening has a direct influence on the degree of complexity it introduces. An opening at a ridge is very simple to deal with, because the rain water drains away from it. An opening close to the eaves will interrupt the flow of drainage from all the upslope area; this may be difficult to weather, particularly if the roof surface is profiled.

Concealed-fix roofing is often installed in very long lengths. Certain commercial systems have special details which allow thermal expansion to occur, without creating significant thermal stresses; these include sliding-on clips, moving clips and slotted holes. If the roof opening is to accommodate a rigid penetration, such as a chimney, there must be an allowance for the sheets to move, usually involving an oversize kerb surround, and a flexible flashing.

As explained elsewhere in this book, there has been a recent trend towards multi-layer roofing incorporating vapour checks and breather membranes. A vapour check is only effective when it is fully sealed, so it must be carefully sealed around any opening. Breather membranes are intended to provide a clear run for draining condensate; this is not possible at a roof opening, and it must be accepted that the value of a breather membrane may be greatly reduced around roof openings.

The roof covering is supported by some form of structure (battens, purlins, continuous deck, etc.), but the structure does not normally continue through an opening. It will usually be necessary to include structural *trimmers* around an opening. This can require extremely accurate workmanship; if the structural trimmers are built into the roof structure, any creep in fixing the roof covering may mean that the opening fails to coincide with its supports. Sometimes the trimmers are adjustable so that the opening may be made to coincide with the 'as laid' roof sheets.

Tiles and slates are used on relatively steep slopes, and openings may be incorporated without great difficulty. The most common large openings in tiled roofs are at chimney stacks. When these are located at the ridge the weathering is very simple, but when they occur at mid-slope a little more thought is required. The solution is to form a *soaker* around the opening. A *soaker* is a specially prepared sheet which matches the roof profile, for the purpose of overlap joints, and incorporates a raised kerb around the opening.

Figure 11.1 shows how such a soaker may be formed, on site, from sheet lead. The lead is easily formed to match any tile profile, and is embedded in a mortar joint to ensure its seal to the brickwork. The levels are arranged so that stormwater flows from the tiles onto the lead, behind the chimney stack. The lead-lined gutter, so formed, is higher than the flashings at the sides of the stack; hence the water discharges freely at either end. At the bottom of the stack it is sufficient to simply lap the lead over the tiles, and gravity will ensure a satisfactory run-off.

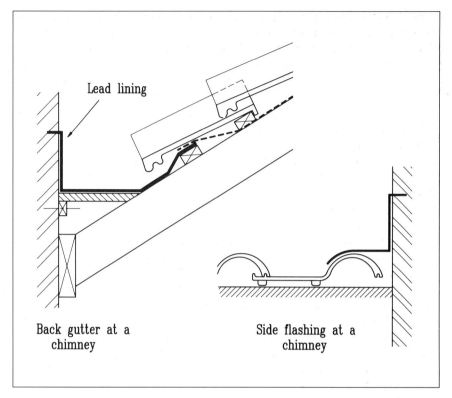

Back gutter at a
chimney

Side flashing at a
chimney

Figure 11.1 Lead flashings around a chimney stack.

Small openings for vent pipes are formed in several ways, two of which are shown in Figure 11.2. The traditional method was to use a *lead slate*, in place of one tile or slate. A lead sleeve was soldered into a piece of flat lead sheet, and this was fitted into the roof by site forming the lead base where necessary.

Lead slates are still common, but there is now a move towards simpler methods which require less craftsmanship. One such solution is the plastic tile incorporating the vent pipe. These are available for many of the popular tile profiles, and in a range of colours which enables them to match most clay and concrete tiles. The vent pipe is usually adjustable in the tile so as to accommodate any roof slope.

Other special tiles are available to accommodate small ventilator outlets. These may simply provide some natural ventilation, or they may form the outlets of forced ventilation systems. In either case, the standard component is much simpler than a soldered lead fabrication, and is more pleasing aesthetically.

Profiled sheets are often laid to shallow pitches. The sheets are larger than tiles, and a different approach is required to the weathering of openings.

The preferred arrangement for small openings in profiled sheets is the flexible soaker. This consists of a cone of synthetic rubber, with a base incorporating a soft metal flange. This can be formed around the profiled sheet, and is sealed with gun-applied silicon sealant, and fixed in place by rivets. The top of the cone is cut off to suit the diameter of the pipe. A typical arrangement is shown in Figure 11.2.

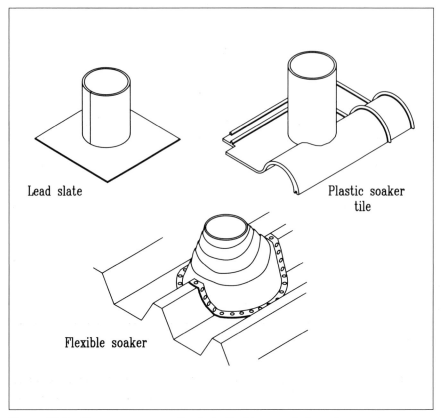

Lead slate

Plastic soaker tile

Flexible soaker

Figure 11.2   Some common methods for flashing small-diameter openings.

Openings in profiled sheet roof can be quite large, however, for instance when they are required for rooflights, ventilators or access hatches. As with any other roof form, the openings are simple when they occur at the ridge, and present varying degrees of difficulty when they are mid slope.

Aluminium profiled sheets may be welded, so it is possible to manufacture soaker sheets, in the factory, with totally weatherproof welded kerbs. They may be painted to match the roof if necessary. Figure 11.3 shows a typical aluminium soaker. At the upslope end, the profiled sheets overlap the flat bed of the soaker, and the ribs are closed by means of foam filler pieces. The profiled sheets must stop short of the kerb to allow stormwater to drain around the sides of the soaker. At the sides, the soaker is formed with ribs to give standard side laps. At the downslope end, the soaker simply overlaps the profiled sheets.

It is not possible to make similar soakers in coated, galvanized mild steel. The heat of welding would burn away the plastic coating and the zinc layer, and it is not possible to restore the protective coating to such fabrications, so they would have very limited durability.

However, such soakers may be produced in glass reinforced plastic; they may even be profiled at the upslope end, and formed to produce a back gutter. These can look

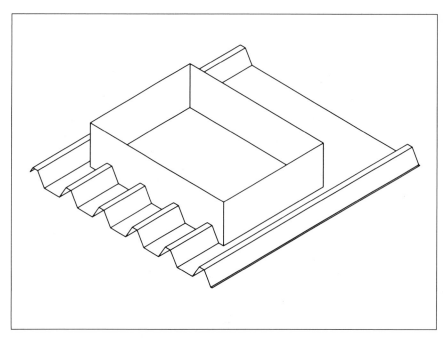

Figure 11.3   A typical factory-made soaker for profiled metal sheets.

extremely attractive initially, but it is common for the GRP to deteriorate, or fade, faster than the plastic-coated steel, and this can detract from the appearance of the roof.

All of these soakers require that the profiled sheets are set out accurately on site. Any significant creep will result in the openings being in the wrong place. In the case of aluminium roofing, it is possible to cut the opening at exactly the right place, and site-weld a kerb around it. This removes the need for exceptionally accurate workmanship. The economics of site welding vary according to the difficulty of access and the number of openings, but site welding has the advantage that the metal skin is continuous, and there is no dependence on mastic sealants.

Another solution to the problem of openings in profiled metal, or fibre cement, roofs is to ensure that water cannot enter those troughs which will be blocked by the kerb to the opening. This is achieved by laying a flat sheet from the upslope kerb to the ridge. The flat sheet must be at least as wide as the opening. This is an inexpensive arrangement, which does not demand special materials or skills. However, it is not attractive to look at, and many architects reject this detail on roofs which are highly visible.

This method is also the easiest way of incorporating an opening into an existing profiled sheet roof. For steel and fibre cement sheets, the alternative involves stripping out areas of roof sheets to install a soaker. In the case of aluminium roofs, it is possible to site-weld soaker units. These have an attractive appearance, and are secure against leaks; however, they may prove to be expensive by comparison with alternative methods.

When soaker kerbs are formed in thin-gauge steel or aluminium or in GRP, they are intended to keep out the weather — they are not major structural items. Such kerbs may support a plastic domed rooflight, up to about 1200mm square, but they would not be suitable for a heavy access trapdoor, or a large mechanical ventilator. Heavy items should be mounted on structural supports immediately inside the kerb of the soaker. If thermal movement of the roof sheets is expected, there should be a small clearance between the structural support and the soaker kerb.

It is not possible to be precise as to the maximum size of openings in profiled sheet roofs. There is no limit to the length in the direction of the slope, the opening could extend from the ridge to the eaves. However, the width depends on the amount of stormwater which is to flow in the trough at either side of the opening. A sheet without openings would normally have only a few millimetres of water in each trough, even under severe storm conditions. At a soaker, the water from a number of troughs is redirected to the first trough on either side of the opening. If this trough is totally filled, the side laps or crown fixings may become submerged, and this will inevitably lead to leaks.

Of course the size of the troughs varies between different profiles, and the amount of water in the troughs is influenced by several variables. Steep pitches drain quickly, thus reducing the depth of water. The distance from the ridge to the opening affects the quantity of water to be drained. The severity of the worst anticipated storm varies from site to site, and, of course, the width of the opening affects the amount of water to be re-routed.

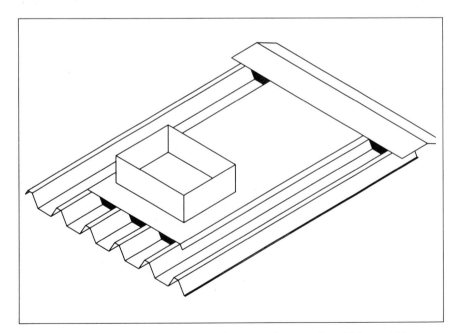

Figure 11.4   Water may be prevented from entering troughs which would be blocked by the soaker kerb.

As a very approximate guide, it may be assumed that openings 1m wide are almost always possible; up to 2m wide will probably be satisfactory, unless the profile is very shallow (say, less than 30mm deep), or the soaker is a long way from the ridge (say, more than 10m). Widths of 3m should only be contemplated for deep profiles, and openings close to the ridge.

At shallow slopes there is a tendency for water to pond behind the soaker kerb. Theoretically the foot of the kerb is horizontal (parallel with the ridge), and water will flow to the open ends and drain away. In practice this is not always the case. The soaker may be slightly deformed, or the roof may have settled slightly; in either case ponding may result. It is also possible for deposits of dirt to form small dams which can contain appreciable amounts of water.

Ponding must be prevented wherever possible. Roof sheets are designed to shed water, not to be permanently submerged. Plastisol coatings on steel are damaged by standing water, and this can lead to premature failure. Even mill-finish aluminium can be damaged because the water contains various pollutants; normally these are so dilute as to be harmless, but evaporation can concentrate the solutions until they become aggressive. Coated aluminium would fare little better, as the coating would be damaged in the same way as plastisol.

The safest course is to prevent the ponding, and this can be done by ensuring that there is a positive fall behind the soaker kerb. Figure 11.5 shows how the base of the soaker may be modified to ensure that efficient drainage takes place.

Sometimes it becomes necessary to locate several openings close together. In such cases it is important to ensure that a reasonable space is left between adjacent kerbs.

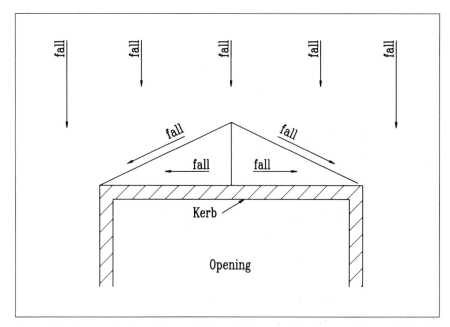

Figure 11.5   In very low pitches, additional falls can prevent ponding.

Drainage channels can be blocked by leaves, straws, dead birds, etc., especially if the pitch is shallow. The absolute minimum width should be regarded as 150mm, and this should be increased wherever possible.

In the case of fully supported metal roofs, openings are fairly straightforward in most cases. The welts and seams, which are used in general roof areas, can also be used around soakers.

The soakers themselves can be formed from flat sheets, with upstands at the kerbs. Figure 11.6 shows how a corner of a lead soaker would be constructed; an extra piece must be soldered in, to complete the corner. Lead has very little mechanical strength, so the lead kerb is simply weather protection to the structural kerb, which is often formed in timber. (Lead has been used here as an example, the comments are equally applicable to any metal used as fully supported roofing.)

Built-up felt, or single-membrane, roofs are the easiest of all in which to form openings. The surface is free from seams or profile, and there are no drainage channels to redirect. The nature of the product, and its application, makes for ease of construction. Figure 10.5, in the previous chapter, shows some typical kerbs for openings, and the importance of timber fillets at upstands was described in Chapter 6.

If openings are simple to construct, the designer may become complacent, with disastrous results. Felt and single-membrane roofs are often used at extremely shallow pitches, and such roofs drain very slowly. A little settlement, compression of the insulation, or deflection of the deck can result in ponding, and standing water will seek out any weakness in the membrane. The drainage can be improved by creating extra falls by the method shown in Figure 11.5.

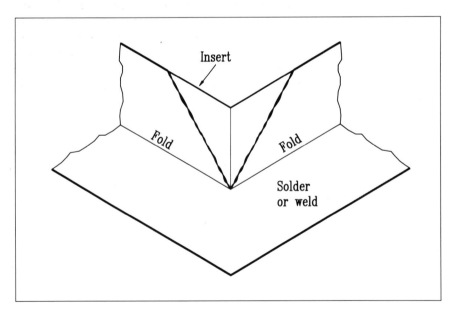

Figure 11.6   Lead soakers may be formed on site by cutting, bending and soldering.

It is also important to leave sufficient space between adjacent openings. As mentioned earlier in this chapter, 150mm is the absolute minimum width for a drainage channel.

When the roof lining is visible, as is the case in most industrial and commercial roofing, the opening must be aesthetically pleasing from the inside as well as weathertight from the outside. In insulated roofing the kerbs must be insulated to prevent excessive heat loss, and to minimize the risk of condensation. Figure 10.5, in the previous chapter, shows a range of typical kerb types which meet these requirements.

It has already been stated that breather membranes lose much of their effectiveness at openings, because the continuous drainage run, for condensate, is disrupted. It is therefore of paramount importance that the vapour check is securely sealed around the opening — if the vapour check is a true vapour barrier, then the breather membrane is virtually redundant. This requirement must be communicated on the building site in the strongest possible terms.

Most of this chapter has been concerned with the design of roof openings in new construction. These are not necessarily simple, but the designer has several alternatives available, and is able to ensure that the roof specification can accommodate openings. The forming of openings in an existing roof, however, can be more complicated, especially if the original roof design was made without thought of the possibility of openings. It is not possible to give specific advice as to how to deal with every application, but the general principles outlined in this chapter should provide a useful starting point in most cases.

Many buildings are constructed as speculative ventures, to be leased after they have been completed. It is quite possible that the leaseholder may require a roof opening for some reason specific to his business (perhaps for an extractor fan over a paint booth, or a flue over a furnace). Some far-sighted architects anticipate this possibility by including one or more soakers in the roof of speculative units. These are simply fitted with lids which may be removed if a roof opening becomes necessary. In this way the openings become new build rather than modifications to existing work.

FURTHER READING
The Lead Sheet Manual: A guide to good building practice - The Lead Development Association.
Trade literature.

12   FLASHINGS

Plate 12.1   Interlocking tiles with mortar-bedded ridges and hips, a stepped lead
flashing at the chimney, and a lead-lined valley gutter.
(By courtesy of Redland Roof Tiles Ltd.)

Most roofs, except for domes and barrel vaults, consist of one or more inclined
planes. Each plane has a high point, usually the ridge or a hip, and a low point, usually
the eaves or a valley. There are two ends, and these may be verges or hips.

Roofs made of tiles, profiled sheets, fully supported metal or membranes, could
continue indefinitely in a single plane; it is at the edges of a plane that some form of
special attention is needed. This special attention usually takes the form of a *flashing*.

A flashing provides the transition from one roof plane to another, as at a ridge or
hip; or from a roof plane to a vertical plane, as at a verge or mono-pitch ridge. The
primary purpose of the flashing is to maintain a weathertight joint at such intersects.

The flashing is sometimes used to achieve an aesthetic effect. This may be done
by using flashings in colours which contrast with the roof sheets, or in a completely
different material, or a special shape.

The designer should always remember that the main purpose of a flashing is to
provide long-term weather protection; the function of the flashing should not be
compromised in the cause of aesthetic effect. However, it is usually the case that steep

roof pitches are highly visible, and shallow roof pitches are not. Steep roof pitches drain much more efficiently than shallow ones, hence they are more easily rendered weathertight, so it is possible to pay more attention to the appearance of those flashings which are most likely to be seen.

A further function of a flashing is that of accommodating tolerances. A roof structure is subject to building tolerances, and the roofing elements are subject to manufacturing tolerances, and fixing tolerances. A good flashing detail will take account of the effect of the accumulated tolerances, and will make good any discrepancies or irregularities.

There is no limit to the number of angles, shapes and configurations at which flashings may be required. In this chapter the main roof types will be discussed in terms of ridge and verge flashings; unless special mention is made, it may be assumed that a hip is a sloping ridge. Many manufacturers of roofing products publish product manuals which show how their products should be flashed.

Tiled roofs are relatively steeply pitched, and are usually visible; it is important that the flashings look right for the roof, and that they have similar durability to that of the roof.

The traditional approach was the *mortar bedded* ridge, sometimes known as a *wet* ridge. A typical arrangement is shown in Figure 12.1. A half-round ridge tile is produced in the same material as the roof tiles (i.e. concrete or clay), and this is firmly bedded in a sand/cement mortar. Tile slips are normally used to bridge the gap between the two slopes; this helps to prevent the mortar from slumping. The ridge tiles are butt jointed, and the joints are sealed with mortar.

The mortar must fill any gaps between the ridge tile and the roof tiles; this is easily achieved when using plain tiles, but is more difficult with profiled tiles. In fact the usual approach for profiled tiles is to use *dentil slips*; these are narrow, rectangular strips of tile material which are bedded in mortar in the troughs of the profile. This creates a much more level bearing surface for the ridge tile, and the dentil slips close the large gaps at the profile troughs.

The corresponding verge detail is also shown in Figure 12.1. An *undercloak* of plain tile is used as a soffit; this is fixed so as to slope slightly away from the wall, thus helping to prevent water from running back into the joint. The tiles and undercloak should project about 40–50mm beyond the face of the wall or barge board. An overhanging tile at a gable is particularly vulnerable to wind gusts and turbulence; such tiles require extra support, and this is provided in the form of *verge clips* which are securely nailed to the tiling battens. The verge clips are exposed to the weather, and must be made from a material which possesses both strength and durability; preferably stainless steel. A sand/cement mortar is used to close the space between the roof tiles and the undercloak. The space must be fully filled if the detail is to be weathertight.

An alternative modern arrangement is the use of *dry* details. These are flashing arrangements which achieve a weathertight joint without mortar. Examples are shown in Figure 12.1. At the ridge, a half-round ridge tile is secured, by means of stainless steel fasteners, into a suitably positioned timber support. The ridge tile is seated on proprietary UPVC strips; these may be shaped to match the contours of

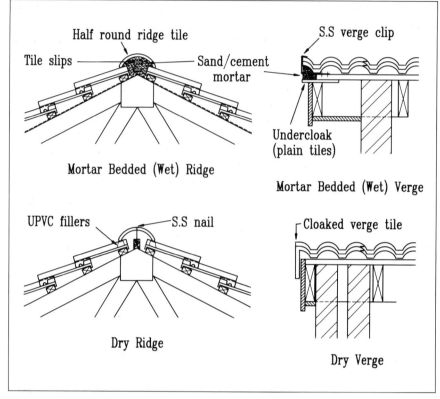

Figure 12.1   Alternative 'wet' or 'dry' details for tiled roofs.

profiled tiles, or straight for use with plain tiles. UPVC jointing straps achieve weathertight joints between successive ridge tiles.

At the verge it is possible to use *cloaked verge* tiles which incorporate their own barge. These tiles are formed with socketed ends to ensure continuity at overlaps. They are usually given a double fixing: they are nailed to the tiling battens, and clipped to the tile below.

Figure 12.1 shows a mortar-bedded verge with a barge board, and a dry verge at a gable wall. Both details are suitable in either case, the illustrations simply demonstrate the various possibilities.

Hips can be treated as sloping ridges for all mortar-bedded applications, and for dry applications using plain tiles. A similar procedure is also possible for profiled tiles, but it is not practicable to produce profiled UPVC fillers for every hip angle and roof pitch. Proprietary sealing systems are available, and the advice of the tile manufacturer should be sought.

Slates and tiles have many features in common, and the same may be said of their flashing details. Figure 12.2 shows some of the standard details for slate roofs.

Ridge tiles are available for use with slates. They usually take the form of an inverted 'V' and are produced in two colours: slate grey, to match the roof colour, and orange-red as a contrast; colour choice is largely a matter of personal taste. The ridge tiles are bedded in mortar in a similar way to plain tiles.

Mortar-bedded verges are also very similar to those for tiles, in that an undercloak is used, and the space between the undercloak and slates is filled with a sand/cement mortar. However, it is usually possible to obtain special slates which are 50% wider than normal; this allows the break band pattern to be maintained at the verges without using half-slates. This in turn allows for more secure nailing of the verge slates, so it is seldom necessary to use verge clips.

Another traditional ridge detail for slate roofs is the lead roll. The lead is formed around a shaped timber roll, in just the same way as for fully supported lead roofing. This forms a neat and durable flashing, and weathered lead has a fairly similar appearance to that of slate. Mortar-bedded ridge tiles and lead rolls are both suitable for flashing hips. Proprietary dry-fix systems are also available for use with slates.

Flashings for bituminous shingle roofs are usually made by cutting up some shingles into strips, and forming the strips into flashings (it may be necessary to heat the shingles before attempting to bend them). Verges can be formed in the same way, or special verge trims may be fitted.

Proprietary pressed metal 'tile' roofing systems may have standard flashing arrangements, or may be treated as profiled metal roofs.

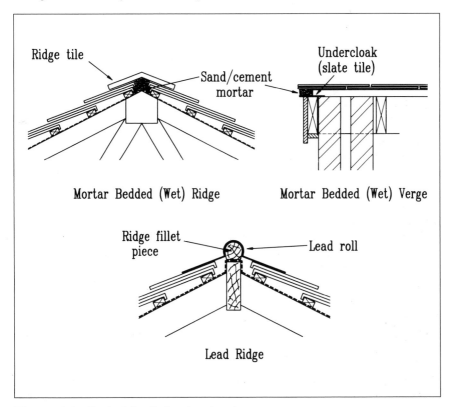

Figure 12.2   Typical details for slated roofs.

Profiled metal roofs are usually fitted with flashings which are purpose-made for the particular roof. The flashing usually matches the roof colour, although a contrasting colour may sometimes be used to produce a special effect.

When the flashings are to match the roof, it is a good idea to make them from the same gauge of metal as the rest of the roof; this should ensure that flashing material is available without the need to paint a coil of a different gauge. Consequently most steel flashings are in 0.7mm gauge, and most aluminium flashings in 0.9mm gauge.

The flashings in Figure 12.3 are typical of what can be produced in metal. In all cases they are shown with welted edges. A welted edge consists of bending about 20mm of metal through 180° along the edge of the flashing. In the case of plastic-coated galvanized steel, this improves durability by protecting the exposed metal at the cut edge. In the case of aluminium, welted edges are useful simply for their stiffening effect.

Ridge flashings are straight, so the troughs of the profile under the flashing must be closed. By far the most common solution is to use profile-cut *foam fillers*. These are formed from a compressible, closed-cell foam material such as polyethylene or foamed EPDM. The action of fixing the flashing compresses the filler and produces a weathertight seal.

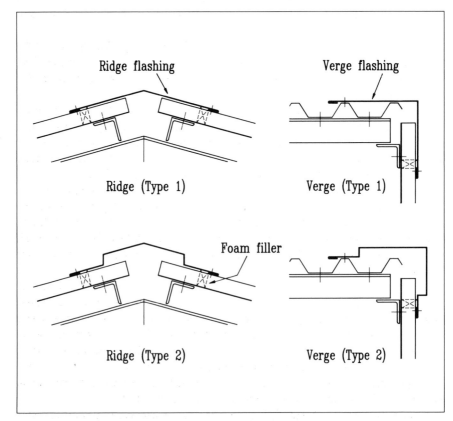

Figure 12.3   Alternative ridge and verge details for profiled metal roofs.

The flashing detail should be designed so that it is very difficult for the filler to become dislodged. Placing the filler behind the welted edge is useful in this respect, as is locating the fasteners between the filler and the edge of the flashing. Most foam materials are eventually degraded by ultraviolet radiation, so the filler should be sheltered from direct sunlight wherever possible. On some sites birds or rodents attack the fillers and destroy them, and in such cases it is necessary to protect the fillers, usually by a toothed or castellated edge to the flashing.

In Figure 12.3, the ridge described as 'Type 1' is the simplest arrangement and also the most economical in terms of both material and labour. This is a perfectly efficient detail from a weathering point of view, but is sometimes thought to lack visual appeal, because such flashings often appear to ripple or undulate, an unavoidable effect in wide strips of thin-gauge metal. The distance from the ridge line to the edge of the flashing is usually 200–250mm, and it is inevitable that 0.7mm steel or 0.9mm aluminium will distort when fitted in such widths.

To overcome this criticism, ridge 'Type 2' evolved. Here the maximum distance between bends or welts does not exceed 150mm, and this greatly reduces any tendency to buckle or ripple. Ridges of this shape present a feature which can add to aesthetic quality of the roof.

The verge flashing must be matched to the ridge flashing because the two flashings will coincide at the gable peak, i.e. a Type 1 verge is used with a Type 1 ridge, and so on.

The vertical leg of the flashing is usually 100–150mm deep. The top edge of the vertical cladding or wall is often uneven because it is site-cut to follow the line of the roof slope. The vertical leg of the verge flashing must be deep enough to cover this uneven edge, and present a neat appearance.

The horizontal leg should be long enough to cover the second profile rib, and is fastened at that rib. This increases the security of the detail, as any water which is driven under the flashing will drain to the eaves by way of the covered trough. A long horizontal leg is also important to accommodate tolerances, as in the course of fixing a large number of roof sheets, it is inevitable that some creep will occur.

Flashings for profiled sheets must usually be fixed to the roof sheets rather than the main structure. Traditionally, rivets were preferred, and are still quite popular. The rivets must be *blind*, in other words they must not allow the mandrel to fall out, leaving a hole. Some rivet designs develop petals or claws behind the jointed metal, this is particularly useful for thin-gauge flashings and profiled sheets, and for aluminium generally.

The greatest criticism of rivets is that they can be slow to install; they require a hole to be pre-drilled, then the rivet is inserted and expanded, and finally a colour cap is fitted. Recently self-drilling, self-tapping screws have grown in popularity. These screws have an integral colour cap, so the installation is reduced to a single operation. The screws have a very coarse thread to enable them to achieve adequate anchorage in the profiled metal sheets. The manufacturer will provide guidance as to the limits of his product but, as a general guide, screws should be treated with great caution in steel of less than 0.7mm gauge, or aluminium of less than 0.9mm.

It is vitally important that the fixings are strong enough to resist any wind loadings to which the flashings may be subjected. Flashings are located at positions where the roof changes shape; it is at these same positions that wind turbulence effects are greatest, and an insecure flashing will be ripped off during a storm.

Hips may be treated in much the same way as ridges, except that the foam fillers must be fitted on the skew. It is possible to buy skew-cut filler pieces, but the supplier will need to know the hip angle on plan, as well as the roof pitch. Most hip flashings intersect with the ridge flashing in some form of 'Y' configuration. Type 1 flashings are much more easily joined at such intersects. There is a temptation to specify factory-made intersection pieces, but this should be resisted; the roof sheets are affected by an accumulation of building tolerances, and it is best to fit the flashings to the roof, as it has been built, and form the intersect on site.

Because the flashings are purpose-made for the individual roof, it is sometimes thought that flashing design is easy. This is not strictly true: as there are several important points to bear in mind in the design of metal flashings.

As was mentioned earlier, metal flashings will ripple or buckle if the distance between successive bends becomes too great. It has been found that for aluminium the distance between bends should not exceed 300 times the thickness of the metal (i.e. 270mm for 0.9mm gauge); for good visual effect, this should be reduced to 150 times the thickness. For steel it is probably reasonable to increase these ratios by 30% but, as steel is often used in thinner gauges, the maximum widths will be about the same. It is usually easier to introduce extra bends, than to change the metal thickness.

Before designing flashings, it is prudent to establish the width of the metal sheets from which the flashings will be cut. The most common coil width is 1220mm, a relic of the days of imperial dimensions. When this width is used, a 610 mm girth flashing will be very economical as there need be no scrap; however, a 615 mm girth flashing will be very wasteful of material. When flashing blanks are cut on a guillotine, material which cannot be used immediately is usually scrapped, it is not economically viable to pay operatives for searching through piles of offcuts.

The shape of the flashings will determine whether they will nest, i.e. fit tightly, one inside the other, in the same way as profiled sheets. This makes for much simpler packing and transport. (The Type 1 flashings in Figure 12.3 will nest, but the Type 2 flashings will not.)

Flashings which nest may be joined by means of a simple overlap, with sealant to complete the weather protection. Flashings which do not nest cannot be overlapped, and are usually joined by means of *butt straps*. A butt strap looks like a very short piece of flashing, but is slightly smaller than the pieces to be joined. Both pieces of flashing overlap the butt strap, and both are sealed. A typical butt strap is 100–150mm wide.

Flashings are commonly supplied in lengths of 3m or 4m. Greater lengths are sometimes possible, and the reduction in the number of joints must increase security; on the other hand, longer lengths are more difficult to handle, and difficulties may result from thermal movement.

It is good practice to make provision for thermal movement when joining aluminium flashings, usually by ensuring that the overlapping flashings do not have fixings through both thicknesses at the overlap. Steel expands and contracts less than aluminium, and it is rarely necessary to make any thermal movement provision in steel flashings.

It is sometimes suggested that the ridge flashing may be replaced by a crimp-curved, floclad ridge sheet, as it is argued that this eliminates the foam fillers and makes an attractive detail. This is to overlook the fact that common site practice is to complete one slope before starting the other (this is economic in scaffolding and guard rails). The curved ridge is only successful if both slopes proceed together; otherwise creep will prevent the ridge sheet from fitting. It should also be appreciated that the use of a curved ridge sheet will require the use of a curved verge flashing at the gable peaks.

Profiled fibre cement sheets have much in common with profiled metal, and could be flashed by means of purpose-made metal flashings. However, there are certain standard items which simplify the design process, and some of these are shown in Figure 12.4. The *close-fitting ridge* is a rather clever method of forming a standard ridge which will fit any pitch of roof. The two pieces interlock and can open and close like a hinge; the pieces form a transition into the roof profile, so foam fillers are not required. Because the flashing is in two pieces, the roof can be built a slope at a time without problems from creep. There is also a fibre cement *eaves closure*. This fitting again eliminates the need for foam fillers.

Roofs covered in continuously supported metal do not usually require flashings as such. The details for wood-cored rolls, batten rolls, standing seams and welts can

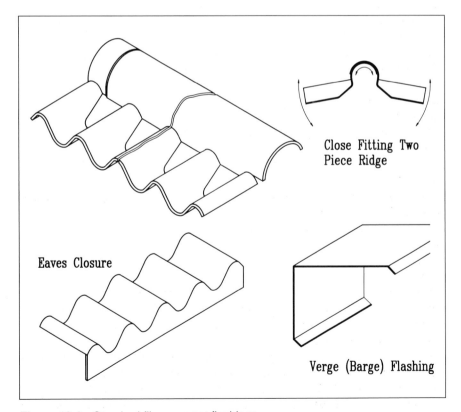

Close Fitting Two Piece Ridge

Eaves Closure

Verge (Barge) Flashing

Figure 12.4 Standard fibre cement flashings.

easily be modified into ridge, verge and hip details. (See Chapter 5 for explanations of these terms.)

Roofs covered in any of the various forms of membranes also have little need of flashings. The membrane is simply fitted around the contours and features of the roof, and the weatherproof membrane is continuous.

However, it is sometimes difficult to terminate the membrane in an attractive manner at the edge of the roof. There are a number of proprietary *fascia trims* to meet

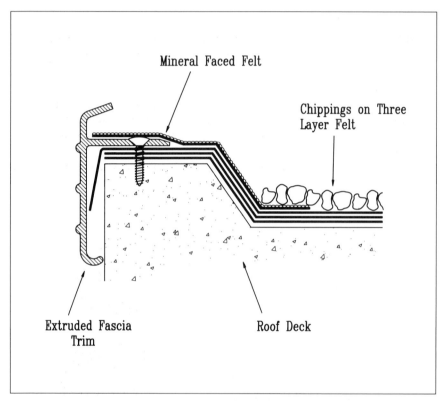

Figure 12.5   An attractive edge detail for a built-up felt roof.

this requirement. A typical arrangement is shown in Figure 12.5; the trim may be extruded aluminium or UPVC. Extruded aluminium may be left as plain, mill-finish metal, or may be anodized or polyester powder-coated. UPVC is formed in self-coloured material. These trims may be used in the same way with single-membrane systems and asphalt. They are of course not really flashings, as they do not contribute to the roof's ability to keep out the weather, and are purely aesthetic.

There is often a requirement to form flashings on site from an easily worked material. Here lead is by far the most popular choice. Lead is also the standard choice for any flashings which are to be dressed in the mortar joints in masonry, e.g. at parapets, stepped roofs, chimney stacks, etc. The only real alternative to lead is high-purity aluminium. This is often very competitively priced, but requires a rather higher standard of craftsmanship than is needed for lead. Aluminium flashings should always be used with aluminium roof sheets, but require a protective coating if they are to be built into mortar joints.

Flashings are a very small part of a roof in terms of area or cost; however, they play a major part in keeping out the weather, and can make a great contribution to the overall appearance of the roof. Time spent in designing flashings is therefore rarely wasted.

STANDARDS
BS 1178:1982 - Specification for milled lead sheet for building purposes.
BS 6561:1985 - Specification for zinc alloy sheet and strip for building.

FURTHER READING
Defect Action Sheet 114 - Slated and tiled pitched roofs: flashings and cavity trays for step and stagger layouts: specification - BRE - 1988
The Lead Sheet Manual: A guide to good building practice - The Lead Development Association.
The David & Charles Manual of Roofing - John H. Wickesham - 1987

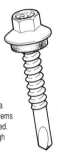

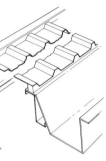

ACCESSORIES

Many of the chapters of this book are concerned with the major items required to construct a roof (tiles, profiled sheets, thermal insulation, rooflights, etc.) These major items are assembled in conjunction with a variety of minor items, none of which justifies a chapter to itself, but whose importance is acknowledged by their inclusion here.

It is almost inevitable that a roof will require some form of *fastener* to anchor it to the supporting structure, or to connect separate elements of the roofing materials. The range of fasteners is vast, as manufacturers have devised products to accommodate most applications. There is not sufficient space here to describe them all, but the more common forms are mentioned, as are several products of particular merit. Most fastener manufacturers publish comprehensive guides to their products, together with advice on preferred fixing procedures.

The simplest fixing of all is the *nail*, the usual fixing for tiles and slates. Plain tiles and slates usually require little more than nails themselves, profiled tiles are usually

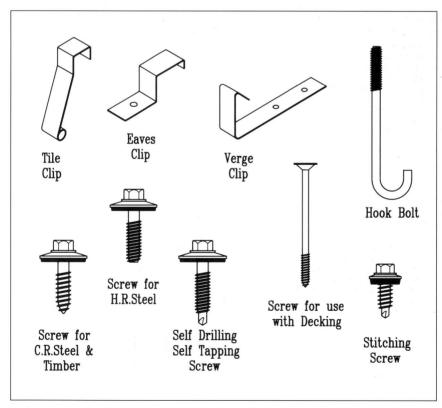

Figure 13.1   Some of the more common types of roof fasteners.

fixed by means of nails in conjunction with purpose-made clips. Some typical clips are shown in Figure 13.1.

It is important to remember that tiles and slates are extremely durable, and are expected to perform their design function for up to 100 years. It is therefore essential to select fasteners which can offer similar durability–the life of the roof should not depend on its cheapest component!

Stainless steel is probably the most suitable material for both nails and clips, as it has high mechanical strength as well as supreme durability. Aluminium and copper are also used, and they too possess excellent durability, but are unlikely to match stainless steel in strength. Galvanized steel nails are no longer used to any great extent because their maximum life is incompatible with that of the tiles.

The size of nails should be determined by reference to the design loading, but 3.35mm is the most common diameter and the length is usually within the range 38–75mm.

The choice of material for nails and clips should take account of any timber preservative which may have been applied to the tiling batten. For example, copper-based preservatives could be aggressive to aluminium, and in damp conditions aluminium nails could corrode and cause premature failure of the roof. Organic treatments, such as creosote, are usually safer than non-organic treatments based on copper and mercury salts.

The usual fixing for profiled fibre cement sheet is the *hook bolt* (see Figure 13.1). These are fitted at the crowns of the profile, and the hook goes around the bottom of the purlin. The type illustrated would be used on an angle purlin, but other shapes are available for use with Z-sections, joists, channels, etc. The seal is achieved by a large compressible washer, and the nut and thread are protected by means of a plastic sleeve fitted on site.

Hook bolts are usually formed in galvanized mild steel; aluminium and stainless steel are possible alternatives where greater corrosion resistance is required.

When profiled metal was first used as a roofing product, it was fixed with hook bolts in the profile crowns in just the same way as fibre cement. However, it was soon realized that the flat trough of trapezoidal profiles could accommodate a sealed washer, provided that it was adequately compressed. *Self-tapping screws* were developed with a coarse thread for timber and thin-gauge purlins, and a much finer thread for thicker-gauge steel (e.g. hot rolled angles and joists). Such screws use less metal than the bulky hook bolts, give a neater appearance, and are easily installed by tightening with a power tool (such treatment would shatter fibre cement).

Screws of this type depend on suitable washers, usually of 16mm diameter for steel roof sheets, and 19mm for aluminium sheets. The size of the washer is partly governed by the need to keep out water, and also by strength considerations. When wind gust attempts to lift the sheet off the roof, a large washer is needed to prevent the fastener being pulled through the sheet. When plastic sheets, are used as rooflights, the washers may be as large as 30mm diameter for this reason.

Screws are produced in mild steel, with a variety of protective coatings, and in stainless steel. A properly protected mild steel screw will probably have sufficient durability for use with steel roof sheets, but it is always sensible to use stainless steel screws with aluminium (bimetallic action is unlikely to cause corrosion in this case, but mild steel screws have been known to promote corrosion of aluminium sheets).

Most screws can be made to match the roof sheets in colour. Some are supplied with colour-coated heads, some with integral coloured plastic heads, and others with separate coloured plastic caps, which are pressed on to the screw heads after fixing.

Screws are now available in a 'self-drilling' form. The screw point drills a pilot hole for the self-tapping screw. These screws increase productivity by reducing the number of separate tasks in installing a fastener. They also contribute to more reliable quality, as it is impossible to select the wrong drill diameter for the pilot hole. However, care should be taken over the selection of the power screwdriver; this should operate at the right speed for the drilling and tapping operations in the purlin. The manufacturer will advise on suitable speeds and appropriate power tools.

When sheets are fixed over 'semi-rigid' substrates such as polyurethane foam insulation board, there is a risk that the insulation could become compressed by repeated applications of snow load, wind suction or foot traffic. This would loosen the screws and allow water to enter at the screw holes, a real problem when polyurethane insulation was first introduced. Manufacturers responded, however, by designing special screws for these applications.

One such screw has become known as the 'stand-off'. It has a short length of very coarse thread immediately under the head; when the screw is tightened, the last two turns cut a thread in the metal roof sheet and hold it securely against the washer. These screws are very successful in steel sheets, but are not always to be recommended for use with thin-gauge aluminium. It would be easy to overtighten the fixing and strip the thread in the aluminium sheet; the aluminium sheets will expand and contract more than steel, and this may elongate the holes and reduce the 'stand off' support. In any case, there is a risk of bimetallic corrosion when two dissimilar metals are in intimate contact. (This risk was even greater when stand-off screws were made in zinc-coated mild steel. Modern stand-off screws can be made from stainless steel, which is much safer in such conditions.)

Other screws have spacer sleeves which pass through the drilled hole in the sheet, and then expand as the screw is tightened so as to support the top sheet. These are safer with thin-gauge aluminium sheets, but the length of the spacer is critical to the success of the fastener; variations in the thickness of the insulation board can reduce the effectiveness of the support.

In metal roofing, the traditional fasteners for side laps and flashings are *rivets*. These rivets are usually made of aluminium, with a central mandrel of mild steel, stainless steel or aluminium. Any rivets in the outer skin must be blind, i.e. it must not be possible for the mandrel to fall out and leave a hole. It is unusual for mild steel mandrels to cause significant corrosion, but they can lead to unsightly rust stains.

The simplest rivets 'POP' rivets, depend on the expansion of the rivet stem by the withdrawal of an oversize mandrel. They have the advantage of low price, but cannot always provide sufficient anchorage for flashings. Greater strength is available from rivets which form 'petals' or 'claws' behind the metal to be joined. Both types of rivet are illustrated in Figure 13.2.

Self-drilling stitching screws are increasingly used in place of rivets. The screws may sometimes cost a little more, but they reduce fixing time on site because there is no need to drill the hole as a separate operation. The stitching screw may be supplied with a coloured head; rivets can be fitted with plastic caps, but this is another site operation. The screws must not be overtightened, or the thread may strip in the roof sheets; however, the screws always tap threads in a double layer of metal, so this is not such a critical condition as was described for stand-off screws.

Many types of membrane roofing require fasteners to anchor the membrane to the structural deck. These may provide anchorage in different ways. The fasteners may hold down the insulation, with the membrane bonded to that insulation. Or the fasteners may hold down the bottom layer of a multi-layer system, with subsequent layers bonded to the bottom layer. Or the fasteners may anchor a single-layer membrane, from a concealed position within the overlap joints.

Steel decks are at least 0.7mm and often 0.9mm gauge; properly designed self-tapping screws can provide adequate pull-out strength in these. A typical screw is shown in Figure 13.1. Similar screws can be used for timber decks provided that there is a reasonable length of thread.

Such screws are less suitable for aluminium decks. A typical deck thickness would be 0.9mm or 1.2mm; this would not be likely to produce a high enough pull-out strength. The recommended alternative is a long rivet which opens up large petals under the deck; the strength of such a rivet is independent of the gauge of the deck. A typical example is shown in Figure 13.2.

Both the screws and the rivets must be used in conjunction with extremely large

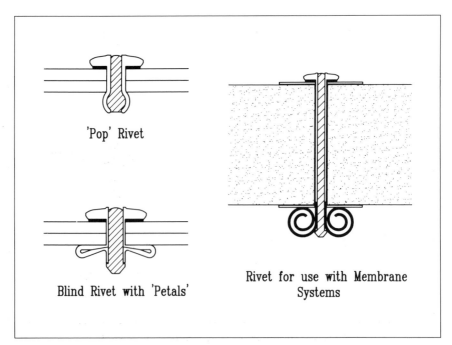

Figure 13.2 There are rivets to meet most fastenener requirements.

washers (perhaps 50mm diameter, or the equivalent in square or rectangular shapes). This is necessary in order to develop sufficient pull-through strength at the insulation board or the membrane.

A long-standing puzzle for roofing designers has been how to secure the weather skin of the roof to the supporting structure without puncturing the vapour check. A recent introduction has been screws with a special coating which seals around the point of penetration. The coating is softened by the heat generated by the friction of the screw against the deck; the softened material accumulates at the interface of the screw and vapour check. On cooling, this material forms a secure seal around the screw. This process is illustrated in Figure 13.3. It is applicable to normal self-tapping screws and to the long ones used with membrane systems.

Many secret-fix systems make use of purpose-designed clips and brackets to hold down the roof sheets. The manufacturers of such systems will normally publish technical literature to explain the design and use of their products.

Some roof forms have no need of further components to keep out the weather. Tiles and slates are usually fixed at relatively steep pitches so they shed rainwater very efficiently, and membranes and fully supported metal roofing provide a virtually unbroken weather skin. This is not the case for profiled sheets; these are used at low pitches, so that laps must be sealed and profiles must be closed at flashings.

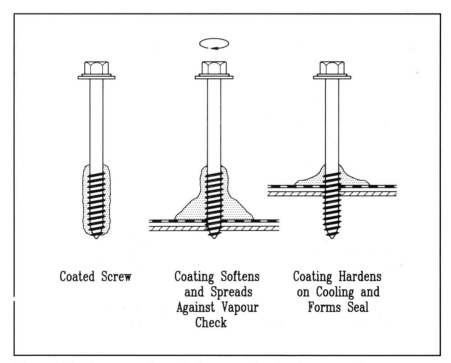

Coated Screw   Coating Softens   Coating Hardens
                    and Spreads   on Cooling and
                    Against Vapour   Forms Seal
                    Check

Figure 13.3   Screws can be sealed to vapour barriers by means of special coatings.

Sealants are available in preformed strips, and in cartridges for gun application. The traditional material for sealants was mastic; today the most durable strip sealants are made from cross-linked butyl mastic, and gun-applied sealants are usually silicon compounds.

Strip sealants are preferred for overlap joints, as they provide a continuous, uniform seal. Gun-applied sealants are most useful where the gaps to be closed have irregular width; they are not usually used for lap joints, although clear silicon is used with translucent rooflights for aesthetic reasons. Silicon sealant is also available in a wide range of colours.

Strip sealants are produced in a range of sizes, the most popular being 9 x 3mm, 12 x 2mm and 4.5mm diameter. The 9 x 3 is preferred for steel sheets, and 12 x 2 for aluminium sheets. This is probably because the greater stiffness of the steel assists in achieving greater compression of the sealer strip. The 4.5mm diameter is easy to compress at first, but becomes progressively more difficult; this can be useful where the gaps are slightly more irregular, or where there are two separate seals to be compressed at the same time (e.g. at the side lap of a composite panel, where there is a second seal at the lining).

Some manufacturers of profiled steel sheets recommend 75mm end laps, completely filled with 75mm-wide sealer strips. The object here is to fill the overlap and thus deny access to rainwater or condensation, which could attack the steel through the thin backing coat. This arrangement is neither necessary nor desirable with aluminium; it would be almost impossible to compress such a wide seal, and two separate strips are a better option.

Profiled fillers are formed from a compressible foam material, which can be clamped in position by a flashing; the filler takes the shape of the profile and makes a good weathertight seal. There are two common materials for foam fillers: polyethylene and EPDM. Polyethylene has the lesser density, and the lower cost; but is degraded by solar radiation and is easily damaged by birds. EPDM is more resistant to ultraviolet radiation, and its greater density also enables it to resist attacks by birds.

EPDM is usually black, and this can be a disadvantage when it is used in conjunction with pale-coloured sheets. A partial remedy is to set the filler well back under the flashing where it is in deep shade (this also protects the material from exposure to solar radiation). Some manufacturers produce polyethylene fillers in two colours, one half black and the other white; the user chooses which side is to be on view. Another possibility is to use EPDM facings on polyethylene fillers. This gives good resistance to solar radiation and bird attack, at lower cost than a solid EPDM filler.

Many profiled sheets are asymmetric, because some details require the filler to be under the sheet (e.g. eaves), whereas at other positions it may be over the sheet (e.g. ridge). The fillers should be specified as large rib or small rib. The difference is indicated in Figure 13.4.

At hips and valleys, skew-cut fillers are needed; these are supplied to order by most manufacturers. The calculation of the angle of cut involves both the hip angle and the roof pitch; the geometry is quite complicated and is best left to the supplier.

There are also various versions of ventilated fillers. These are described in more

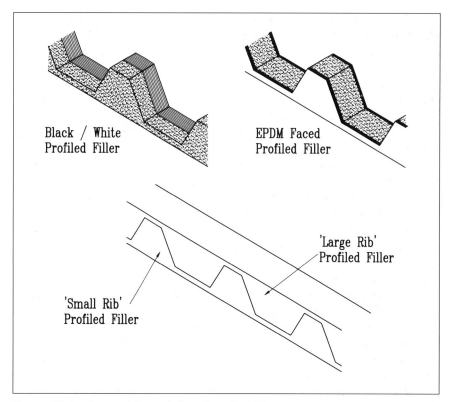

Figure 13.4   Some of the varieties of profiled fillers.

detail in Chapter 9. The best arrangement is usually to have the ventilation openings at the top of the ridge fillers (to reduce the risk of water entering), and at the bottom of the eaves fillers (to allow condensate to drain out).

Sandwich construction roofs, with profiled outer and inner skins enclosing mineral fibre insulation, require rigid *spacers* to prevent compression of the insulation. These can be in timber, but the most popular product is the galvanized Z-section — these are cheaper than timber, do not require such long screws, and will not shrink or warp due to heat or dampness.

These spacers are good conductors of heat, and can seriously reduce the insulation value of the cladding. However, the Building Regulations permit designers to ignore their effect when calculating the U value of the roof. It is good practice to use a heat barrier between the outer skin and the flange of the spacer; compressed mineral wool 10mm or 12mm thick is suitable. The heat barrier maintains the insulation value; it also keeps the spacer warm and thus reduces the risk of its becoming wet through condensation.

Liner sheets are made from thin-gauge steel or aluminium. The ribs are easily crushed, so it is not usually possible to mount the spacers directly over them. A popular solution is to place the spacers on plastic packs or ferrules; these are situated in the troughs of the liner, and are just tall enough to lift the spacer clear of the ribs.

The ferrules must be strong enough to transmit all loads to the purlins (the design case is probably foot traffic, with the weight of a man directly over a ferrule). It is not really true as is sometimes claimed that plastic spacers act as heat barriers The underside of the spacer is in the warm part of the roof, the heat barrier must be on the outside of the spacer.

The arrangement of a Z-spacer over a plastic ferrule is not very strong in torsion; thermal movement of the sheets can twist the spacer and bend the fixing screws. Some systems use metal brackets to create a more rigid support for the spacers; for these it is quite clear that some form of heat barrier is needed.

Insulation boards and plasterboards are often supported in *T-bars*. These can be in plastic or aluminium, but the usual material is galvanized mild steel. It is possible to use these in natural finish, but they do not look very attractive, especially as they grow older and spots of rust appear. When they are used to support white-faced boards, the visible face of the T-bar can be painted white, or covered with a white tape.

There are many special accessories for use with *tiles*. These include ventilation grilles for fitting at the eaves, ventilator outlets for the ridge, tiles with integral ventilators or vent pipe openings, transparent tiles, etc. Most of these items are formed in plastic, and visible parts are coloured to match the roof. Such items are often specifically designed for a particular tile profile, and tile manufacturers issue trade literature describing their products.

Proprietary *ventilators* are available for use with membrane roofing. These small items are fitted in the membrane at regular intervals, releasing vapour which would otherwise be trapped and could cause blistering. Manufacturers of such items should be consulted as to their installation and use.

Knowledge of *adhesives* has advanced rapidly in recent years. Products, in tape and liquid form, are now available which can provide satisfactory long-term performance on exposed roofing. The application instructions should be studied and followed carefully. For example, it may be necessary to clean and dry the parts to be bonded, and to work above a specified temperature. There is no doubt that some adhesives are becoming more 'user friendly', and it is certain that adhesives will become more widely used in the future.

STANDARDS
BS 1494:Part 1:1964 - Fixings for sheet, roof and wall coverings.
BS 5889:1989 - Specification for one-part gun grade silicone sealants.
BS 8000:Part 6:1990 - Code of practice for slating and tiling of roofs.

FURTHER READING
Trade literature.

ACOUSTIC PROPERTIES

As environmental awareness has grown, noise has been identified as a form of environmental pollution. This has led to increased demand for some form of acoustic performance from roofing materials.

Unfortunately, acoustics are not always well understood by either architects or builders; consequently some attempted acoustic solutions are doomed to failure. The purpose of this chapter is to explain the basic acoustic properties and definitions in order that the reader might progress to more advanced works, if necessary.

There is no single measure of acoustic performance, but four distinct properties each of which is a form of acoustic performance. The obvious starting point is a definition of these different properties; this will set the scene for everything which follows.

It is often necessary to impede the passage of sound, for example to prevent the noise from aircraft disturbing guests at an airport hotel, or to prevent noise escaping

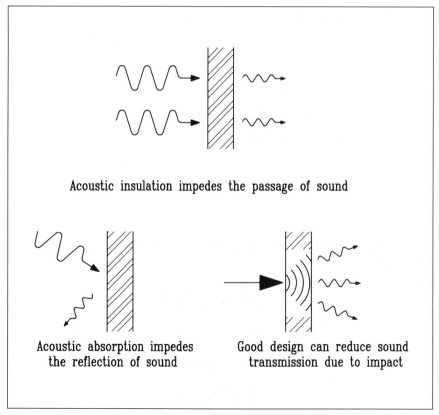

Acoustic insulation impedes the passage of sound

Acoustic absorption impedes
the reflection of sound

Good design can reduce sound
transmission due to impact

Figure 14.1　A diagrammatic representation of three different acoustic properties.

from a discotheque, and annoying local residents. In either case the requirement is for *acoustic insulation.*

One reason for confusion in acoustic matters is a result of using the word 'insulation' without qualification. It is sensible to take a disciplined approach and refer to 'acoustic insulation' or 'thermal insulation', as appropriate. This becomes especially important when using a product such as mineral fibre which can be used in either type of insulation, although it is not equally efficient in the two applications. There are other thermal insulants which make very little contribution to acoustic insulation.

In an auditorium, or lecture theatre, it is important that sounds can be heard clearly. Echoes which blur the sounds cannot be tolerated. Suitable surfaces (usually soft) are therefore introduced at the floor, walls and roof. These surfaces do not easily reflect sound, and they are said to provide *acoustic absorption.*

Acoustic absorption becomes increasingly important as the size of the room increases; sports halls and swimming pool buildings are often enclosed by hard surfaces, and are notorious for 'poor acoustics'.

A design to give acoustic insulation will not necessarily provide acoustic absorption, however. The two requirements must be considered independently.

Some other acoustic problems associated with roofs can be avoided at the design stage. Heavy rain or hail will cause *drumming* on a roof. This is most likely to be severe on a metal roof, and thus to be a problem if the roof is directly over sleeping accommodation. The effect can be minimized by careful attention to design details, and the choice of a suitable acoustic insulation material. Some roofs are prone to *creak* as a result of building movements. Such noises can be a nuisance in theatres, classrooms, or sleeping accommodation. The exact mechanism of this source of noise is not properly understood, but the materials most likely to cause it are known, and can therefore be avoided.

In order to make sensible decisions about acoustic performance requirements of roofs, it is first necessary to understand some basic facts about the physics of sound. Sound is the perceived effect when the eardrum detects minute pressure variations in the air. The source of the pressure variations will usually be a vibrating solid body (a hammer, a musical instrument, a machine, etc.) in contact with the air.

The sound travels through the air as a pressure wave. This is analogous to a ripple on a pond; the ripple spreads from the original source simply by a vertical rise and fall of the water surface, and there is no movement of water across the pond. Similarly, sound travels by pressure variations in the air, there is no movement of the air.

The further a ripple travels from the source, the longer it becomes; as it becomes longer it also becomes shallower. This is simply a case of its energy becoming more widely dispersed. In just the same way, sound energy becomes dispersed as the distance increases. Figure 14.2 shows how the area, influenced by the sound, increases with distance; the sound is weakened in proportion to the area.

Weakening or reducing a sound is a simple enough concept, but it does not become useful until the sound can be described in measurable units. Many builders and

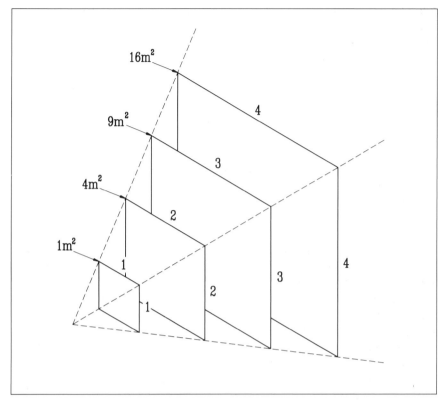

Figure 14.2   Sound intensity reduces with the square of the distance from the source.

architects have difficulty in understanding acoustics because the units are peculiar, and do not behave like other units!

First, it is important to realize that sound is an effect perceived when pressure variations are detected by an eardrum. It could be argued that unless there is a listener, there can be no sound! That debate may be left to the philosophers, but it is the interaction between pressure waves and human perception which is used to establish the system of acoustic units.

As the scale of units is clearly concerned with sounds which can be heard, the zero point is taken as the lowest level of sound which can be detected by the average ear. This lowest level is generally taken to be represented by a power output of 10^{-12} W/m^2. This then becomes the zero point on the scale; it is sometimes described as 'the threshold of audibility'.

It is vitally important, for everything which follows, that this is fully understood. Zero sound is the least sound that a normal person can detect, it is not an absolute absence of sound. The zero point is said to have a sound pressure level of 0 bels. A sound pressure ten times as great is said to be 1 bel, a further ten times as great, 2 bels, etc.

Table 14.1   Typical sound levels from everyday experience.

Source of sound	Sound level (dB)	Perceived effect
Complete silence	0	Totally inaudible
Ticking of a watch	20	Extremely quiet
Suburban garden	40	Quiet
Conversation at 2m	60	Loud
Shouting at 1m	80	Very loud
Typical discotheque	100	Extremely loud
Pneumatic hamme at 1m	120	Intolerable
Aircraft engine at 4m	140	Intolerable

The 'threshold of pain' is said to be 12 bels because a sound of such a high power can actually damage the ear (12 bels is a million million times more powerful than 0 bels). It is becoming apparent that the use of bels keeps the numbers reasonably simple. In fact bels are rather too coarse to be really useful as units. A finer unit is achieved by dividing a bel into ten divisions, each of these is a decibel (written as dB).

Table 14.1 gives an indication of the power of various sounds within the experience of most listeners. These cannot be highly accurate as no two watches, gardens, conversations, etc, will ever be identical. However, the table provides a useful guide, and places the units in a familiar context.

It is worth considering how several sounds may be added together to determine their combined effect. It is tempting to refer to Table 14.1 and say that five ticking watches, at 20 dB each, are equal to a discotheque at 100 dB. This is quite clearly nonsense, so some other approach is needed. The correct method is to use logarithms, taking the sound level in bels as being a logarithm. So 1 is the logarithm of 10, 2 is the logarithm of 100, etc. As an example, take the addition of two 60 dB sounds:

$$60 \text{ decibels} = 6 \text{ bels}$$
$$6 = \log (1{,}000{,}000)$$
$$6 \text{ bels} + 6 \text{ bels} = \log (2 \times 1{,}000{,}000)$$
$$= 6.3 \text{ bels}$$
$$= 63 \text{ decibels}.$$

It is seldom necessary to make such a calculation, but an understanding of the principle involved is useful, and helps to prevent any absurd answers being calculated by inappropriate use of units.

It is now possible to investigate acoustic insulation in a little more depth. It has been stated that sound is transmitted through the air in the form of pressure waves, and that these originate from a vibrating body. Now suppose that the pressure wave arrives at a wall or roof; although the pressure variation is extremely small, it actually causes the wall or roof to vibrate. This, in turn, causes the air on the other side to vibrate, and hence the sound is transmitted onwards. As might be imagined, heavy walls or roofs vibrate less easily than light ones. This is the basis for acoustic insulation, whose most important property is mass.

However, there is a further complication; acoustic insulation is not equally effective at all frequencies. It is more difficult to insulate against low-frequency sounds (e.g. deep notes) than against high-frequency sounds (e.g. high notes).

There is an approximate mathematical formula to calculate acoustic insulation values:-

Sound reduction (dB) = [20 x log (M x F)] - 43, where M is the mass of the roof in kg/m^2 and F is frequency of sound waves in hertz

This formula may be used to calculate a series of values ready for future use. The ear is sensitive to a wide range of frequencies; usually taken as 20 Hz to 30 000 Hz. In the case of acoustic insulation it is usual to quote figures from 63 Hz to 4000 Hz, or sometimes 8000 Hz. It is not necessary to go further as insulation levels will be good at high frequency.

The formula can be used to produce acoustic insulation values for a range of frequencies, and for different weights of cladding. This has been done, and the results listed in Table 14.2. This table provides useful approximate guidance, but nothing more. The formula has its limitations, and acoustic insulation is not solely governed by mass. The introduction of an air cavity within the roof construction can improve the acoustic insulation by up to about 5dB; but the air gap must be at least 50mm, and most authorities believe that 100mm is necessary for an appreciable benefit.

There is also the possibility that a roof, or roof panel, may have a resonant frequency which coincides with the frequency of the sound. When this happens, the acoustic insulation is very significantly reduced. In practice this means that most roofs have a frequency at which their acoustic insulation is relatively poor.

The formula is at its most useful when the frequency of the sound is known. However, it is more usual to be faced with the general question of sound insulation for an unknown range of frequencies. In these cases it is normal to take the acoustic insulation in the 500–1000Hz range, as being a useful basis for comparison. Often the mean of these is used.

There is some discussion as to whether the weight of the purlins should be included

Table 14.2 Approximate sound reduction, in decibels, for roofs of different masses, at a range of frequencies.

Mass	Frequency Hz							
Kg/m^2	63	125	250	500	1000	2000	4000	8000
5	7	13	19	25	31	37	43	49
10	13	19	25	31	37	43	49	55
20	19	25	31	37	43	49	55	61
40	25	31	37	43	49	55	61	67
80	31	37	43	49	55	61	67	73
160	37	43	49	55	61	67	73	79

with the weight of the roofing. It is usually included on the grounds that the roof can only vibrate freely when the purlins are also vibrating. The contribution from light cold-rolled purlins would probably be about $2kg/m^2$.

Single-skin aluminium on light cold-rolled purlins would weigh $5kg/m^2$. Double-skin insulated aluminium, on cold-rolled purlins, would weigh around $8kg/m^2$, and the substitution of steel for aluminium would increase this figure to about 14. Similarly, a steel deck with rigid insulation board and a single membrane weather protection, would weigh around $20kg/m^2$ including cold-rolled purlins; substitution of three-layer felt and chippings would increase the figure to about 45.

Tiles or slates, with sarking, battens, underlays, etc. could have a weight in the region of $40–80kg/m^2$. Concrete roof slabs, with asphalt finishes, can weigh considerably more.

Simply adding more weight to improve the acoustic insulation can be an expensive option; all the roof weight must be supported by the structure, and extra weight will often result in increased structural costs. It may be more economical to introduce cavities.

The most common house designs have a pitched tiled roof, over a loft space, which is separated from the living accommodation by a plastered ceiling. The heavy roof, large 'cavity', and relatively heavy lining, meet the acoustic insulation requirements for all but the most noisy locations. Of course, this would not be a practical solution for a large clear-span factory building.

In terms of acoustic insulation, the units are simpler to apply; if the acoustic insulation of a roof is 30dB, then the sound power will be 30dB less on one side than on the other. A 30dB reduction is achieved by reducing the sound energy by a factor of ten, three times; so the sound energy is reduced, in all, by a factor of 1000.

It should be appreciated that these comments apply to continuous roofing details. Poor construction, for instance with gaps, allows sound to find alternative entry points. Similarly, light plastic rooflights, in a heavy roof construction, may have a dramatic effect in reducing the value of acoustic insulation.

After acoustic insulation, the next most important acoustic property is acoustic absorption. When sound, i.e. a pressure wave, arrives at a roof or wall, some of it is reflected; the amount reflected is governed by the acoustic absorption of the surface.

Usually, hard surfaces have low absorption, and soft surfaces have high absorption. There is an obvious analogy with a rubber ball being thrown at a wall; it bounces back off a hard wall, but not off a hanging curtain.

Absorption is usually measured in terms of the *absorption coefficient*. This is the fraction of sound energy absorbed when falling upon a surface. It follows that a small coefficient applies to a surface which absorbs little and reflects much, i.e. a hard surface. In the same way, a large coefficient applies to a surface which absorbs a lot, and reflects little, i.e. a soft surface. The coefficient is a fraction; it cannot be less than zero, nor can it be greater than one.

The coefficient can only be established by testing in controlled conditions. As with acoustic insulation, the results vary with the frequency of the sound; figures are usually tabulated to show the effect of frequency. Table 14.3 shows a selection of typical values; floor and wall figures are included in order that example calculations

Table 14.3 Typical sound absorption coefficients, for a selection of roof linings, at a range of frequencies.

Material	Frequency Hz					
	125	250	500	1000	2000	4000
Plaster ceiling with airspace behind	0.30	0.15	0.10	0.05	0.04	0.05
Softwood boards on joists	0.15	0.20	0.10	0.10	0.10	0.10
25mm polystyrene with cavity behind	0.10	0.25	0.55	0.20	0.10	0.15
Plasterboard in T-bars	0.30	0.20	0.10	0.07	0.04	0.04
50mm woodwool slabs with 25mm airspace behind	0.30	0.40	0.50	0.85	0.50	0.65
Light profiled metal lining with 80mm quilt behind	0.90	0.50	0.25	0.20	0.15	0.15
Perforated metal lining with 80mm quilt behind	0.80	1.00	0.85	0.50	0.45	0.30
Glass as single glazing	0.30	0.20	0.10	0.05	0.05	0.05
Plastic tiles on concrete flooring	0.02	0.04	0.05	0.05	0.10	0.05
Painted brickwork	0.01	0.01	0.02	0.02	0.02	0.03

may be demonstrated. More comprehensive lists are available from other sources.

To properly understand the table it is necessary to understand the mechanics of sound absorption. Soft, fibrous materials absorb sound energy as friction between fibres; open-textured surfaces allow sound to enter, to be absorbed in soft backing material; light panels vibrate and absorb considerable amounts of energy around resonant frequencies.

Plasterboard, metal lining and glazing all have good absorption at low frequency, but are less useful at high frequencies. The introduction of perforations into the metal linings extends the range of frequencies over which there is an acoustic absorption benefit.

The values quoted in the table are given for general guidance only. The absorption properties vary with variations in the application details. Whenever acoustic absorption performance is important, test data should be used as the basis of calculations. Some manufacturers may have this data readily available, but in other cases the tests must be carried out to obtain the necessary figures.

The importance of these properties becomes greater as the size of a room, or hall, increases. Sound can bounce around an enclosed space for a considerable time; the rate at which it decays will depend on the absorption of the various surfaces, and the distance it travels between surfaces. In a small room the reflections will occur a very short time intervals, and at each reflection there will be some loss of energy; if the energy loss is relatively high, the sound will decay extremely quickly.

It should be noted that the effect illustrated in Figure 14.2 applies to outdoor conditions; indoors, there is clearly a limiting area to which the sound can be applied.

Figure 14.3 shows a plan view of a single room. The room contains a sound source at A and a listener at B. Various routes are shown, by which sound can travel from A to B. Route 1 is the obvious direct route, a straight line from A to B. Routes 2 and 3 require a reflection from a wall, route 4 requires two reflections, and route 5 has three reflections. There are many other possibilities in the two dimensions drawn and of course there are vastly more possibilities when the third dimension is included, but the diagram is enough to make the point.

Sound travels at approximately 345m/s, at all frequencies. Suppose the building in Figure 14.3 is about 50m long; route 1 is about 15m, sound will travel along it in about 0.04 seconds. Route 5 is about 100m long, sound will travel along it in about 0.29 seconds. So a word travelling along route 5 will not arrive at the listener until a quarter of a second after he heard it via the direct route. In the meantime, he has received more words by the direct route, and various others as reflections from all the different routes. If the reflected words are not damped out quickly, the effect at B is simply a jumble of blurred sound. This is what is usually meant by the term 'poor acoustics'.

The foregoing comments have identified the cause and nature of a potential problem. What is now required is a means by which the problem can be quantified, and possible remedies evaluated. Fortunately this is a well documented procedure.

The *reverberation time* of a room may be defined as the time taken for a sound to decay by 60dB. Reverberation time may be calculated according to the formula:-

Reverberation time = 0.16 x V/A

where V is the volume of the room in m^3,

and A is the sum of the areas of all surfaces, multiplied by their respective absorption coefficients.

As a general guide, it is usually accepted that the reverberation time should not exceed one second in rooms where speech must be heard clearly. This is usually increased to 1.5 seconds for music.

The reverberation time of the room may be influenced by the floor, wall and roof surfaces. It is often the case that the roof is very easily influenced at the time of construction, but expensive to change later. It is therefore important that the acoustic performance is considered at the design stage.

An example may help to clarify the calculation. A sports hall is 40m long by 25m wide and 10m high; it has a concrete floor with thermoplastic tiles, walls of painted brick, and a roof lining consisting of plasterboards in T-bars.

The reverberation time will be calculated for a frequency of 500 Hz. It is usual to

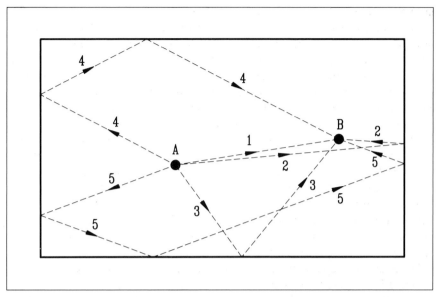

Figure 14.3 Reverberation is the result of sound travelling from *A* to *B* by an infinite number of different routes.

check values from 500 to 2000Hz for general purposes.

The volume is 40 x 25 x 10 = 10000 m^3

Roof area is 40 x 25 = 1000 m^2 1000 x 0.1 = 100

Floor area is 40 x 25 = 1000 m^2 1000 x 0.05 = 50

Wall area is 130 x 10 = 1300 m^2 1300 x 0.02 = 26

So reverberation time   = 0.16 x 10000/(100 + 50 + 26)

                     = 9.1 seconds.

This is so far above the recommended figure that some acoustic problems are inevitable. The floor and walls do not offer much scope for variation, but the roof could be changed to include a lightweight, perforated metal lining. This would change the reverberation time to 1.73 seconds which, although greater than 1 second, is a very significant improvement.

People are efficient absorbers of sound, and are effective over a wide range of frequencies. A person is equivalent to 0.4m^2 of fully absorbent material, so 50 people in the sports hall would increase *A* by about 20, and this would reduce the reverberation time to about 1.69 seconds (a very minor improvement).

The only remaining possible improvement would be to change the design of the walls. If woodwool were to be used for the upper 5m of wall height the reverberation time would fall to about 1.25 seconds, with 50 occupants. This may be considered acceptable for a sports hall, but the exercise has shown how materials can have a major effect on acoustic performance. The calculation should be repeated for 1000 Hz and 2000 Hz in order to achieve confidence in the design.

Woodwool slabs, which have an open-textured surface, are good absorbers of sound, and are also good acoustic insulators by virtue of their reasonably high mass. Similarly, mineral wool is a good absorber of sound, particularly at medium and high frequencies; in its denser forms it is also a good sound insulator.

It should also be appreciated that the formula for reverberation time can only be approximate as it takes no account of the building shape. Figure 14.4 shows how a normal pitched roof tends to reflect sound back into the centre of the room, but a butterfly roof throws the sound to the sides so that it is further reduced before returning to the centre.

Drumming occurs when heavy rain or hail falls on a roof. Each tiny impact generates a sound, and the combined effect of these sounds can be a loud noise. The extent to which the noise is heard inside the building depends on the nature of the roof materials, and on the details of construction.

Drumming is seldom a problem for normal domestic roofing. The outer tiled roof is fairly heavy, there is an enclosed loft space and a ceiling, so sound is greatly reduced before it penetrates the accommodation areas. Lightweight metal roofs over factories or warehouses can be very noisy, however, as the sound has a 'hard' path through the spacers and fixing screws. (This is rather like saying that the cold bridges are also acoustic bridges, and there is some truth in this.) Drumming can probably be tolerated occasionally in factories, but can be minimized at the design stage by reducing the number of 'hard' paths. The use of soft insulation is one method: soft material between the outer sheet and the spacer has acoustic benefits, as well as reducing cold bridges.

If a building has a long reverberation time, then drumming is far more likely to be a nuisance. Suppose that the sports hall in the earlier example had actually been built with a reverberation time of 9 seconds. Any drumming rain sounds would have built

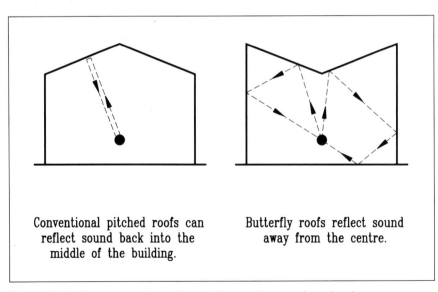

Conventional pitched roofs can reflect sound back into the middle of the building.

Butterfly roofs reflect sound away from the centre.

Figure 14.4 The shape of a roof can influence the reverberation time.

up as new noise superimposed on existing noise. Reducing the reverberation time would clearly reduce the volume of noise from drumming.

Creaking in roofs can sometimes be due to the relief of stresses and strains in the supporting structure, but these are seldom sufficiently loud or frequent as to constitute a nuisance. However, there are a few recorded cases of much more severe noises occurring in lightweight roofs.

The exact cause of such noises is not clear, but they are associated with metal roofing over foil-backed foam insulation boards. It is believed that the noise results from friction between the metal sheets and the foil, but such noises have proved difficult to reproduce in a laboratory. It is possible that the solution is simply to separate the metal from the foil by some means, e.g. the introduction of a breather membrane.

In common with drumming, creaking effects may be amplified by rooms with long Reverberation Times.

Acoustic problems are not easily solved by calculation. However, an understanding of the basic principles can often be enough to ensure that problems are averted. In other cases the early identification of a potential problem can lead to testing, revision of details, or other remedial action.

STANDARDS
BS 3638:1987 - Method for measurement of sound absorption in a
   reverberating room.
BS 8233:1987 - Code of practice for sound insulation and noise reduction
   for buildings.

FURTHER READING
Trade literature.

FIRESTOPPING TO ROOF VOIDS WHERE A VENTILATION GAP IS REQUIRED

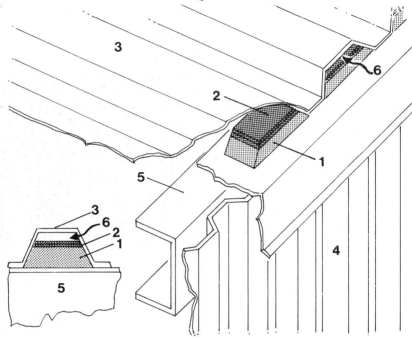

KEY TO TYPICAL INSTALLATION

1. COROFIL VENTED FIRE STOP BLOCK
2. LAYERS OF COROFIL INTUMESCENT STRIP

3. ROOFING PROFILE
4. METAL CLADDING
5. STEEL EAVES BEAM
6. VENTILATION GAP

COROFIL FIRE STOP PRODUCTS

C500 Fire Stop Strip
C260 Fire Stop Blocks
Intumescent Pipe Collars
CF Fire Protection System
Intumescent Firewall System

C144 Firewall & Cavity Barrier
Intumescent Pipe Bands
Fire Stop Compound
CW Fire Protection System

COROFIL FIRE STOP PRODUCTS DIVISION
Pre-Formed Components Limited
Davis Road, Chessington, Surrey KT9 ITU
Telephone: 081-391 0533 Telex: 888711 Corfil G
Fax. No. 081-391 2723

COROFIL is the registered trade mark of Pre-Formed Components Limited.

FIRE

15

Plate 15.1   Fire stops made a major contribution towards containing a fire in one section of this building. (By courtesy of Pre-formed Components Ltd.)

One of the first ways in which mankind was distinguished from pre-human ancestors, was in the use of fire. For thousands of years, fire was mankind's source of heat, light and energy; it helped human beings to survive, and made possible many of their greatest achievements.

However, mankind has paid a heavy price for this association with fire. Fire can kill people and animals, destroy buildings and lay waste vast areas of forest or crops. It can kill by burning, by exhausting supplies of oxygen, or by creating toxic fumes.

Man had little choice but to introduce fire into his buildings, and therefore had to learn how to control fire and to minimize the risks. In Chapter 1 we saw that the building of thatched roofs in London was prohibited by law almost 800 years ago.This was a case of learning from experience, and acting to prevent repetition. Today the Building Regulations contain many requirements in respect of fire, and most of these were first introduced as a response to some particular incident.

Part B of the Building Regulations is concerned with all aspects of fire safety in the design of buildings. The regulations seek to minimise the risk of death or injury to the occupants of the building, the occupants of neighbouring buildings, and innocent passers-by. They include measures to contain fires by means of walls and bulkheads, by the use of non-flammable materials, and by demanding certain details of construction.

The regulations also have specific requirements about rapid means of escape in the event of fire. There are rules governing fire doors and fire escapes, as well as protection of staircases to lengthen the time for which they will remain safe after the outbreak of fire.

This chapter deals only with those aspects of fire, and its prevention, which directly concern roofing. It is worth noting that fire is unique in building design in that provision is made for something which should never occur. A roof is designed to support snow and resist wind, it will experience summer sun and winter frost, and be subject to torrential rain. It will be designed to accommodate all of these, and the design case will be a statistical prediction of the worst event which may be anticipated during the life of the building. There is a fervent wish that the roof will escape any contact with fire; however, it will be constructed to provide appropriate resistance to fire, in case an accident may occur.

A vast number of materials are used in the building of roofs. All of these have different characteristics. It is possible to define requirements for strength in terms of stress at failure, or to measure properties in terms of thermal or acoustic insulation. It is also possible to assess durability by experiment, and from experience. The properties in relation to fire are more difficult to assess.

It is easy to state that a clay tile will not burn, and neither will a profiled steel sheet. However, the clay tile may be fixed on timber battens over an underlay of bituminous felt or plastic. Similarly, the steel sheet may be fixed over a polyurethane foam insulation board. These built-up systems obviously pose some inherent risk, as they include combustible materials; some form of consistent method is needed to make comparisons and assessments possible.

One aid in arriving at such a method is the use of British Standards. These set out test methods with clear criteria for what constitutes a pass or failure. Any such tests are simply a means of comparison, there is no such thing as a British Standard fire! British Standard BS 476 is concerned with a variety of fire tests on building materials. Not all of these involve roofing.

The first important consideration is that of how the outside surface of a roof will respond if subjected to sparks or flames — a building's owner may go to extreme lengths to prevent the outbreak of fire within his own building, but he will have little influence on the behaviour of his neighbours. The law against the use of thatch in London, passed in 1212, was intended to prevent the spread of fire by sparks and burning brands. Figure 15.1 shows how fires are propagated if roofs are made from material which can easily catch fire.

BS 476: Part 3 sets out a test method to be applied to any material which may be used as the outer skin of the roof. The test consists of applying a gas flame to the roofing specimen, in a controlled and defined manner. According to the performance of the specimen, the material is awarded a designation made up of two characters. The first character denotes how well the material resisted penetration; A, B and C are the three possibilities, with A showing the greatest fire resistance.

The second character is concerned with the spread of flame from the point of application; A, B, C and D are the classes available, where A has no spread of flame and the other classes have a maximum defined amount.

Figure 15.1 Combustible roofing materials may be ignited by sparks or brands from neighbouring buildings.

Table 15.1 compares the properties of many popular roofing materials. A number of these are designated AA; this does not suggest that they behave identically when exposed to fire, but it does indicate that they are sufficiently safe for their differences to be unimportant when considered in terms of risk of being ignited by an external source.

The regulations impose restrictions on the proximity to a site boundary at which the various designations may be used. For example, a DA designated roofing material is only allowed to be used at a distance of over 20m from the site boundary, this decreases to 6m for designation AD, and there is no restriction for designation AA (a material with an AA designation is not likely to be ignited by sparks or burning brands). There are further, special rules relating to glass and plastic, and to the physical size of the building.

Just as the surface of a roof must help to resist external fires, so the lining should not contribute to the spread of fire within the building. BS 476: Part 7 specifies a test method to determine the extent to which flame may spread, when it impinges upon various lining systems. Surfaces are classified from 1 to 4 according to the test results; Class 1 is the highest rating.

However, the Building Regulations also introduce a Class Zero surface, and this is a higher rating than Class 1. To achieve Class Zero rating a material must be non-combustible, and this is established by Part 4 of BS 476. (This part of the standard

Table15.1 Typical fire ratings of a range of roofing materials.

	ROOF DESIGNATION	CLASS OF SURFACE	COMBUSTIBILITY	IGNITABILITY
Profiled metal	AA	O	non.com.	P
Fully supported metal	AA	n.a.	non.com.	P
Fibre cement	AA	O	non.com.	P
Tiles	AA	n.a.	non.com.	P
Slates	AA	n.a.	non.com.	P
Glass	AA	O	non.com.	P
Translucent plastic rooflights	AA to AC	0 to 3	com.	P/X
Profiled fibre	AA to AC	n.a.	com.	P
Bituminous shingles	AA	n.a.	com.	P
Bituminous felt	AB to CC	n.a.	com.	X
Mineral-faced bituminous felt	AA	n.a.	com.	P
Foil-faced insulation board	n.a.	1	com.	P
Foil-faced plasterboard	n.a.	O	non.com.	P

sets a test system by which all materials fall into one of two possible categories, namely 'combustible' and 'non-combustible'.)

Within dwellings, linings must usually be at least Class 1, but Class Zero is required over common areas within blocks of flats. The regulations set minimum standards according to the type of occupation, and size of roof (or room). Again there are special conditions for plastic rooflights.

Combustible materials vary widely in the ease with which they may be ignited. It is clearly better to have a roofing material which is difficult to ignite, even if it eventually burns as strongly as an easily ignited alternative. BS 476: Part 5 gives test methods for ignitability; materials are designated X (easily ignited) or P (not easily ignited).

During the 1970s there were several tragic incidents involving loss of life when smoke and fumes spread through voids or cavities to other rooms which were not directly affected by the fire. The Building Regulations were amended to introduce the concept of *cavity barriers* and *fire stops*.

A cavity barrier is closure to a void between a ceiling and a roof, or within the ventilated cavity in a multi-layer roof. It must fill the gap completely, have adequate fire resistance, be stable in shape (i.e. not prone to warp or shrink), and sufficiently durable to provide many years of security without maintenance.

Cavity barriers are used to isolate separate dwellings within a block of flats, or to provide additional security to escape routes in the event of a fire, it is imperative that the stairs should be kept clear of fumes and smoke, and cavity barriers can prevent the staircase from filling with smoke via cavities in the roof or wall construction.

London for many years had special regulations. One of these required that when buildings shared a common wall (e.g. terraced houses), then that wall had to project a considerable distance above the roof, to act as a fire break. This was certainly an effective precaution against the spread of fire, but weather protection was complicated by the need to flash the roof at the party wall, from either side. A more modern approach would be to make the most effective continuous roof, and then use cavity barriers to prevent the spread of fire or fumes beyond the party wall.

There are many examples in London of roofs that have been spoilt aesthetically by walls projecting through mansards or hipped roofs. Of course it is right to put safety ahead of aesthetics, but it is interesting to note that no other British city found this precaution necessary.

The Building Regulations also require that large roofs have cavity barriers at not more than 20m intervals. This is sometimes thought to be unnecessary in large, single-storey warehouses or factories. It is argued that the purpose of the cavity barrier is to prevent a person being overcome by fumes when he is unaware of a nearby fire hazard — a person in a single room should never be unaware of a fire hazard in the same room! The counter-argument states that a building may be subdivided to accommodate a change of use, and it is simpler to install the cavity barriers during construction. In any case, a ventilated cavity may provide a draught to fan the flames and thus contribute to the fire.

Chapter 9, in dealing with the prevention of condensation, mentioned the benefits of providing a clear, ventilated cavity within the roof construction. It has been argued that cavity barriers restrict ventilation, and that they increase the risk of the insulation becoming saturated. The argument continues along the lines that the fire may never happen, but condensation is unavoidable on very cold nights; it seems wrong to cater for an improbable event in a way which worsens a predictable event. This is not an entirely logical argument; it is imperative that fire risk should be minimized, if this creates a condensation risk then that risk must be reduced by other means.

One possible solution is to use intumescent, ventilated *fire stops*. These are fire stops formed from a special type of material which foams and expands when heated.

The fire stops do not completely fill the cavity when they are fitted; they allow a flow of air to provide ventilation. They may quite possibly remain in this state for the entire life of the building. However if there should be a fire, the intumescent material foams and expands until it fills the cavity; the foam then resists the passage of heat or smoke. The intumescent fire stop only closes the cavity when a fire occurs, but this is not reversible; the fire stops must be replaced after the fire or the cavities will no longer be ventilated.

Intumescent materials are also available in the form of paints. A coat of such paint can be a good protection against fire; the coating foams on being heated, and the foam insulates the surface which was coated. This can be an expensive form of protection, but it weighs very little and does not occupy much space. Its usefulness is limited in roofing applications, but it is used extensively on oil rigs and in other high-risk applications. It can provide a means of protecting existing roofing.

From an early age children are taught that, if they discover a fire, they should close the doors and windows to starve the fire of air, then raise the alarm. This is sensible advice for rooms in a dwelling, or for individual classrooms or offices; it may not be applicable to large factories, warehouses, supermarkets or leisure centres. The difference between the two cases is that, in the case of the small room, it is taken for granted that there are no people in the room that is to be sealed off. This is not usually the case with a building which is, effectively, a single large room.

When fire breaks out in a large building, the greatest risk to the occupants is that of suffocation. It is usually possible to retreat from the fire quickly enough to avoid suffering burns, but it is vital to find an exit before the building fills with smoke. People lose their sense of direction very easily when they cannot see, and blunder into obstructions until they are overcome by fumes and lose consciousness.

This particular hazard may be reduced by the use of *smoke ventilators* or *fire ventilators*. These are roof-mounted ventilators which can be used in the normal manner, but which open automatically when the temperature rises to a certain level. The ventilators release the smoke and fumes, and this results in improved visibility within the building. Occupants are thus given a little more time in which they can see their escape route, and rather better air to breathe while making their escape.

The idea of ventilating the roof appears to conflict with the fire safety lesson about keeping the fire enclosed, but preventing loss of life is the most important consideration. The extra air may fan the fire, and this may result in additional damage to property; this must be accepted as being of secondary importance, as property can be repaired or replaced. In any case, the release of heat may reduce the risk of structural damage.

Smoke ventilation can also be achieved in other ways. Certain translucent rooflight materials have a low melting point, and disintegrate on heating, leaving a clear opening. It is important that such lights should not allow burning drips to fall, or contribute significantly to the fire. PVC rooflights have a suitably low melting point, and these were the first 'self-venting' rooflights. The more recent development of polycarbonate lights has provided an alternative to PVC. The polycarbonate has a slightly higher melting point, but compensates for this with greater strength, improved durability and better light transmission.

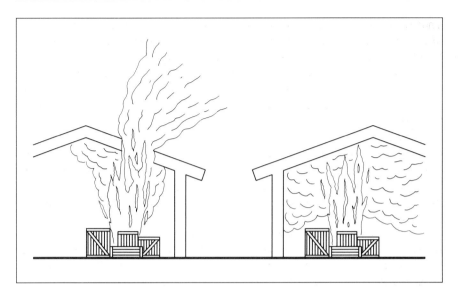

Figure 15.2 Self-venting roofs can release smoke and fumes, and improve the prospect of escape for any occupants.

Profiled aluminium provides yet another form of self-venting roofing. Aluminium alloys lose strength when heated; in the case of the alloys used for roof sheets, the onset of loss of strength occurs at around 100°C. This is very convenient for roofing applications, as roof temperatures are unlikely to exceed that level anywhere in the world. It is also useful for self-venting roofing.

When fire breaks out, the temperature of the roof immediately above the fire rises rapidly. If the roof is aluminium, it loses strength in this local area. Ultimately, the temperature becomes so high that the metal loses its strength to the extent that it is no longer able to support its own weight over the span between the purlins; it collapses, leaving a hole directly above the fire. This is a very effective self-venting mechanism as it applies to any position in the building; smoke ventilators and plastic rooflights are scattered over the roof, and they can only be successful when the fire is close to them.

Any form of smoke ventilation must help people to escape from the building. It also helps the firefighters to locate the fire and concentrate their efforts in the right place. It is easy to imagine the difference from the point of view of firemen arriving at the scene. In the case of a vented roof, there will be a plume of smoke showing exactly where the fire is located, the heat inside the building will be less intense, and visibility may not be totally impaired. By comparison, a non-vented roof gives no clues as to the location of the fire, and the building interior is very hot and filled with smoke.

The Building Regulations make distinctions between types of buildings: for example, dwellings require different treatment from that given to warehouses. However, there is no distinction between different uses within a building type; for example the Building Regulations would not require different treatment for warehouses according to their contents. This may seem an inconsistent attitude, but the regulations apply to new buildings; the use of a warehouse may change during its life, and some contents may be highly inflammable, others less so.

Similarly, the regulations impose restrictions on building materials according to the distance from the site boundary. They cannot take account of what occurs beyond that boundary, because the use of the land may change.

Of course, the building owner should take a responsible view as to the safety of his business, and hazards which may result from the actions of his neighbours. He will be forced into this by his insurance company, the insurer may insist on certain precautions being taken before a policy is issued. Businesses are also subject to inspection by the local fire officer to ensure that escape routes are clear, and that fire fighting appliances are in place.

One extra precaution could be to install a *sprinkler system*. This is a series of pipes and spray heads fitted to the underside of the roof or ceiling. The system is activated by a rise in temperature; the sprays are switched on, and the entire building is saturated. This is very effective in extinguishing fire, but is not without some drawbacks. Sprinklers can be switched on by accident, and the entire stock in a warehouse could be ruined. Similarly, a small fire could activate the system, and the sprinklers could cause more damage to the building contents than the fire would have done. The various risks must be properly assessed. The decision to install sprinklers needs to be made at the structural design stage, as their weight is significant and must be included in the design of the purlins and rafters.

When a risk can be identified, it should be addressed. A store for paint, solvents, gas bottles, or other flammable materials, should be isolated in some way, and perhaps have a fireproof suspended ceiling, or even concrete slabs.

Fires cannot be eliminated. Regrettably there will always be accidents, irresponsible behaviour, and even arson. A building designer should assume that the building will experience at least one fire during its lifetime; if the fire occurs, it has been anticipated in a safe design - if not, no harm has been done.

STANDARDS
The Building Regulations Part B2 - Internal fire spread (linings).
Part B4 - External fire spread.
BS 476 - Fire tests on building materials and structures.

FURTHER READING
Trade literature.

ROOF LOADING

16

Roofing materials are subjected to various forces. These forces must be adequately supported, or the occupants and contents of the building will be put at risk.

The design process consists of a series of procedures which lead to a safe design. First, any possible source of loading or force must be identified. Second, the probable magnitude of any such forces must be estimated. Third, the structural support system must be determined. Finally, the roofing elements must be shown to have sufficient strength to support the loads in the proposed manner.

The ever-present loading condition is that of *dead weight*, or self weight. Every roofing material has some weight, and the weight acts vertically downwards under the influence of gravity. The weight of the material is usually stated in the product literature issued by the manufacturer or supplier.

The dead weight must be supported, but the support may come from a separate component of the roof, as in the case of membranes; or the product may be self-supporting, as in the case of profiled sheets.

It is unusual to design for dead weight in isolation; there are several other sources of loading, and it is almost inevitable that one or more of these will act in the same direction as the dead load. Some combination of forces acting in the same direction is likely to yield the design case.

The other regular contributors of roof loading are snow, wind, foot traffic and thermal effects. Snow may be uniformly distributed, or drifted. Wind turbulence at edges and corners can cause high local loads. People walking on roofs create high point loads. External roofing materials experience the whole range of outdoor temperatures; they expand and contract, and this can cause quite high forces.

Other sources of loading can include the weight of plant, for example, ventilators, supported on the roof; or accumulations of waste material from the processes in the building, such as piles of sawdust on the roof of a sawmill.

Designers would not want to have to work from first principles in order to estimate the loading due to snow for each new building. This would require vast amounts of work, and would create a design dilemma between caution and competitiveness. Fortunately this is not necessary, as there is a comprehensive British Standard which provides design data which is accepted throughout the construction industry.

The British Standard is BS 6399: Part 3: 1988, the Code of Practice for Imposed Roof Loads. The standard was produced from analysis of extensive weather records. It enables the designer to derive the loading condition for any site in Britain. Prior to the introduction of this standard, the usual design method was to take a uniform snow load for the design of any roof in Britain. In effect, the design case was a uniform covering of snow, a little more than half a metre thick. This was revised after several buildings suffered severe roof damage from drifting snow, even to the extent of structural collapse. The whole question of snow loading was investigated by the Building Research Establishment, and the British Standard is based on the findings of the BRE.

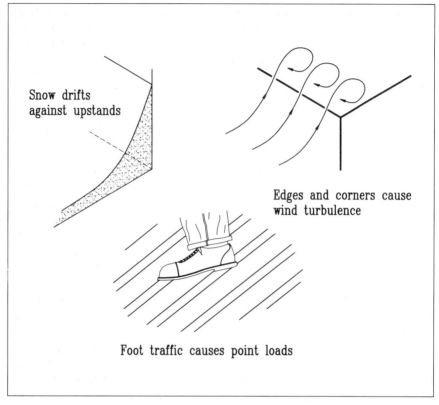

Snow drifts
against upstands

Edges and corners cause
wind turbulence

Foot traffic causes point loads

Figure 16.1   Roof loading can result from snow, wind and foot traffic.

The calculation is broken down into simple stages. First, the basic snow load is established by reference to a map, which indicates the load variations by means of isopleths in $0.1kN/m^2$ increments. Second, the figure is adjusted for sites which are more than 100m above sea level. Finally, the loading is multiplied by a shape factor appropriate to the roof; the shape factors are given in the standard.

In the event of the snow loading being calculated as less than $0.6kN/m^2$ the figure of 0.6 should be used. This is to ensure that a design case cannot be unrealistically low. (Note that BS 5502 allows the use of lighter loads for farm buildings, and this acknowledges the lower risk associated with such buildings.)

The significance of these figures is more easily appreciated in the context of familiar buildings. A typical house has a roof area of about $50m^2$; if the loading is $0.6kN/m^2$ then the roof is carrying three tonnes of snow. Similarly, a small factory could have a roof area of $1000m^2$; at the loading described, the roof is supporting 60 tonnes of snow.

The standard also points out that snow can be driven by the wind so that it drifts against obstructions. The first calculation establishes a uniform loading over the entire roof, but the standard requires that a further examination is made to see whether this load can be redistributed in an unfavourable way.

Roof features such as parapets can become the location for a snowdrift, in fact any step or sudden change, such as valleys, lean-to roofs, plant rooms, monitors, penthouses, ventilators, etc can promote drifting. Figure 16.2 shows some of the ways in which snow can accumulate into drifts.

The standard provides a method of calculation for determining the maximum loading from snowdrifts according to the height of the obstruction, and the normal distributed loading. There is guidance on calculating the restraining force, that is, the force necessary to prevent the snow from sliding down the roof — this is very valuable when designing snow boards.

The standard is based on meteorological records and on statistical probabilities. The user need not be concerned with the statistical approach unless he requires exceptionally high security, or can accept reduced figures for a temporary structure. For such cases, the standard has an appendix showing how to predict loading to different probabilities.

The snow on a roof will normally be melted by heat escaping from the building and will drain away, or areas of snow may slide down the roof and fall off at the eaves. As was discussed in Chapter 8, however, there is a continuing trend towards the increased use of thermal insulation; this in turn reduces heat losses, and this may result in lower rates of thaw, so that it is possible that the design of more thermally efficient roofs may in fact lead to greater roof loads. Such trends can only be conclusively established after prolonged investigation.

The calculation of snow loading by means of the standard can appear quite complicated. However, it is much simpler than the calculation of wind loading. The weight of snow must act downwards, and snowfalls are relatively uniform over very large areas, whereas wind loading can act upwards, and varies considerably over small distances.

During the 1960s, the Code of Practice for wind loading was very simple and predicted fairly light design loadings. Towards the end of the decade there were several major storms which caused considerable damage. The collapse of the Ferrybridge cooling towers received widespread publicity, and there was also severe damage in Sheffield.

The British Standard for wind loading on buildings was already under revision, and the public outcry resulting from the storm damage accelerated this work. In fact, many experts believed that the new standard (CP3: Chap V: Part 2: 1972, which covers design wind speeds, and the loadings produced by these wind speeds) was unduly pessimistic in predicting wind loads. Certainly the new standard suggested that design loads should be increased by a factor of four or five in some cases, and this was attacked in some quarters as a panic measure. There is no clear right or wrong here. The true design case for a building is the worst storm during the life of that building, and the accuracy of the standard may not be known with certainty for another 30 years or so.

The designer's belief in the standard is largely immaterial. Most specifiers require that the design is based upon it, and Building Regulations approval is usually conditional on demonstrating that the building can support wind loads in accordance with CP3: Chap V.

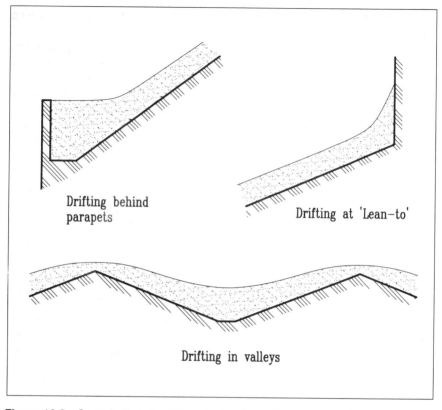

Figure 16.2   Snow is likely to drift against walls and parapets, and at valleys.

The approach taken within the standard is to first establish general conditions, and then to adjust these to take account of special local circumstances. To that extent it has much in common with the Code of Practice for snow loading.

For the purposes of comparison, wind speeds are based on the average speed of a gust of three seconds' duration. Meteorological records were used to establish maximum three-second gust speeds in different locations, at a height of 10m above the ground; these were then analysed statistically. Figures were produced to show, for any location, the three-second gust speed which was unlikely to be exceeded more than once during a 50 year period. It is these gust speeds which form the basis of the wind speed predictions.

A map of Great Britain with wind speeds indicated in the form of isopleths at 2m/sec increments would show the lowest to be around London and in the south-east of England, where speeds can be as low as 38m/sec. In the north of Scotland, the highest speeds approach 54m/sec.

The wind speed for a particular site may need to be modified to take account of special local effects. There will obviously be a difference between the bottom of a valley and the top of a hill, but these two conditions may occur within a few hundred metres of each other, and would not show on a small-scale map of the whole country.

The standard suggests suitable factors to compensate for these topographical variations.

The other local effect concerns the degree of shelter which may be provided by other buildings, trees, hedges or fences. There is an infinite variety of possibilities, but the standard simplifies the analysis by defining four categories of *ground roughness*. Category 1 is flat open countryside, virtually free from any obstructions. Category 2 refers to countryside with trees and hedges, perhaps with farm buildings or small villages. Category 3 includes small towns and the outskirts of large cities; there are many buildings (or trees), but these are typically not more than about 10m high. Category 4 applies to city centres where the buildings are close together, and the average building height is 25m or more.

The standard provides modification factors to adjust the wind speed according to the ground roughness categories.   The selection of the correct category, for a particular roof, calls for extremely careful judgement. The categories may change due to local events; an obvious example is a house built in open countryside, but later surrounded by an industrial estate. The house described would then receive additional shelter, and its design would therefore be conservative. A more serious case would be a building designed to stand in a forest clearing. If the trees were to be felled for timber, then the building would become more exposed, and this could be critical to the safety of the building, unless the possibility was foreseen at the design stage.

There are very few buildings in the areas described by categories 1 and 2, and the areas of city centres, corresponding to category 4, are quite small. It follows that the

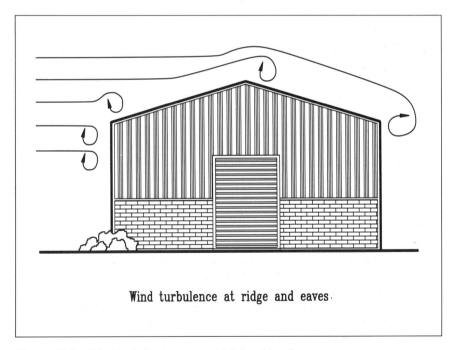

Wind turbulence at ridge and eaves.

Figure 16.3   Wind turbulence causes high local loading.

majority of roofs are located in category 3 areas.

The wind speed is further modified to take account of the building element under consideration, as the size of the element has an influence on the loading which may be applied. Wind is not a continuous flow of air, but a series of gusts; wind loading is a function of the size of the gust, and of the element being loaded.

A suspension bridge must be designed to resist wind loading, but if it is 1000m or more in length, then it is not easily enveloped in a single gust. Its design may be based on the effect of a strong gust over part of its length, or a less severe gust over its entire length. By comparison, a single slate, tile, pane of glass or profiled sheet can be enveloped in a small, high-speed gust.

The standard simplifies the choice to three classes of size. Class A refers to all units of cladding and roofing, and is associated with the highest gust speeds. Classes 2 and 3 are applicable to complete structures, but are not used in the design of roofing or cladding systems. Again, the wind speed is adjusted by means of modification factors given in the standard.

The wind speed to be used in designing a particular building element must be further modified to take account of the height of the element, above ground level. The standard provides further modification factors to include heights up to 200m above ground level.

Of course, it is not possible to state that the design wind speed will never be exceeded; a design can only be based on a statistical probability, itself drawn from previous observations. The standard provides data based on experience, and gives further modification factors to allow for varying periods of exposure. Most roofs are designed for "the worst storm in fifty years", but higher security is sometimes required. Similarly, the design case may be reduced for temporary or short-term buildings.

These comments apply to the geographical location of the site itself, its topography and degree of shelter. The only property of the building which is used in assessing the design wind speed is its height. However, the shape of the building also has a major influence on the wind loading. The same aerodynamic effect which enables aeroplanes to fly will attempt to rip the roofs off buildings! For most roof shapes the wind force will be upwards, and is usually described as *suction* or *uplift*. Flat or shallow pitch roofs generate greater suction loads than steep-pitch roofs.

The largest forces occur close to the eaves, verges or ridges as a result of the wind turbulence associated with edges and corners. When a roof has suffered storm damage, it is quite common for the missing tiles, slates, or sheets to have been removed from these high-risk areas around the perimeter of the roof, and these areas are described as the *critical zones*.

If the roofing material, its fasteners, and the purlins are uniform over the entire roof, then it is the conditions in the critical zones which will govern the design case for wind. Another approach is to increase the number of fixings in the critical zones, reduce the support spacings, or use stronger materials; any of these variations increase the complexity of the design, but make more efficient use of the roofing materials.

An additional component of wind loading results from openings in the building, such as doors or windows. If a door is open during a storm, then the pressure inside the building may be either increased or reduced; this exerts loading on to the roofing and cladding, and can increase the overall loading. For example, a roof deck may be loaded by suction forces from wind blowing over the roof; if, at the same time, there is an increase of pressure inside the building, then the overall upward load on the deck, and its fixings, is increased.

Some multi-layer systems support internal loadings independently of the external loadings. These would include metal profiled sheets as sandwich systems, or timber-boarded sarking under slates or tiles. The designer must remain alert to all loading cases, and to their point of application.

Table 16.1 provides some indication of the magnitude of wind loads, and of their variation. This table is based on a two- storey house with a roof pitched at 20°; it is assumed that doors and windows will be kept closed during a storm, so internal pressures or suctions will be small. The comparison is shown between the four different ground roughness categories, and this is repeated for the London area and the Edinburgh area. The suction figure is the maximum wind uplift loading in the critical zones.

The table shows that an exposed site can suffer approximately double the loading that would be experienced by a sheltered site only a few miles away. Similarly, the wind loading in the Edinburgh area is approximately double that in the London area. Hence, for two identical buildings, one in central London and one on an exposed site near Edinburgh, the wind loads vary by a factor of four. This is a very clear demonstration of the importance of including all the relevant data in the calculation of wind loads.

London and Edinburgh were selected for this exercise because they represent conditions close to the two extremes for Great Britain (the basic wind speed for London is 38m/sec, and that for Edinburgh is 50m/sec). Cardiff and Belfast, by way of example, have a basic wind speed of 45m/sec; this produces loadings approxi-

Table 16.1 The variation of wind loading on the roof of a building, according to the geographical location of the building.

Area	Site description	Suction kN/m²
London	City centre	0.68
	City outskirts	0.92
	Farm outside city limits	1.30
	Airfield outside city limits	1.50
Edinburgh	City centre	1.17
	City outskirts	1.58
	Farm outside city limits	2.25
	Airfield outside city limits	2.61

mately halfway between those for London and Edinburgh.

The standard includes only a limited range of building shapes — pitched roof with central ridge, monopitch, flat roof and multi-span. Any effort to apply it to any other shape will give approximate results at best. It cannot be too strongly emphasized that the analysis of wind loadings is a job for an expert (often a chartered structural engineer), and he may need to carry out wind-tunnel tests in order to arrive at useful answers.

In recent years architects have tended to write vague specification clauses, such as: 'The roofing materials, and their fixings, to be designed to support all loadings in accordance with CP3: Chap V.' This may be despite the fact that the building includes domes, mansards, hips, etc, which are beyond the scope of the standard. In any case the architect is best placed to know the exact site position and its surroundings; he should also be aware of possible future changes around the site.

It is unreasonable of the architect to expect a roofing sub-contractor to take on design responsibilities. Furthermore, it is not in the best interests of his client; competitive tenders should be based on a stipulated set of conditions, not on guesses at interpreting an unclear document, in respect of incomplete information. If wind-tunnel tests are necessary they should be carried out once as part of the design process, not by each prospective subcontractor. This is a very important point; the law is far from clear as to where the responsibility would lie if storm damage resulted from inadequate design by an unqualified person. The architect should seek to ensure that all loading conditions have been fully and properly evaluated.

Wind and snow loading can occur across large areas of roofs; the loading may not be constant, but any changes in loading patterns are likely to be gradual. For example, the depth of a snow drift decreases towards its edges, and the force from wind diminishes away from the centre of a gust. The loads resulting from foot traffic in contrast are totally different in nature. The loads are applied at extremely localised positions, with zero load on all surrounding areas.

The loading from a man walking on a roof includes numerous variables; the weight of the man is of prime importance, but there are other factors to take into account. If the man walks fast, an impact load will increase his effective weight; if he wears soft shoes, there will be no local 'knife-edge' effect, but rigid boots may produce highly concentrated loading. The nature of the roof surface also has an influence; a foot can apply load to a small patch of flat roofing, but cannot wrap around ridges and profiles.

The risk associated with foot traffic must be considered from two distinct viewpoints. The building owner requires that his roof will not be damaged by personnel who have been allowed access to the roof and the personnel require safe access. These criteria may be addressed by providing strong roofing materials, or by limiting roof access.

British Standards 6399: Part 3: 1988 and 5427: 1976 both contain requirements in this respect. It is difficult to walk on roofs which have a pitch greater than $10°$, so the requirements are different for roofs of different pitch. Roofs with a pitch of up to $10°$ can be designed for general access, which means that they may have frequent or substantial foot traffic loading; for instance they could form part of a fire-escape

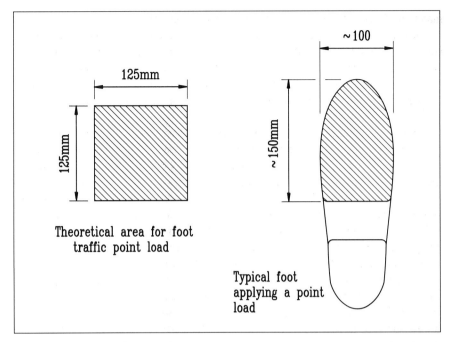

Figure 16.4   No two human feet are alike, but a 125mm square is a reasonable approximation.

system, or provide access to a plant room. Such roofs should be designed for a uniform loading of at least 1.5kN/m².

Roofs of greater than 10° pitch should not be used for general access, and may simply be designed for the calculated snow and wind loads. However, there is always a possibility of point loads incidental to maintenance. The maintenance point load is taken as 0.9kN, applied on a square of 125mm side.

The 125mm square is intended to represent a single foot. Figure 16.4 shows how such a square approximates to the size of the sole of a man's foot. The standards do not require any special treatment for profiled sheets, but most authorities agree that extra attention is necessary. The best recommendation is that the 0.9kN load should be applied to 100mm of a rib of the profiled sheet. This gives a much more concentrated load, as a rib is often 20–30mm wide.

The standards simply require that the sheets can 'support the load', but do not specify a safety factor, or a pass or fail standard. A fibre cement sheet would crack at a certain load, and this is clearly a failure condition. However, a profiled metal sheet may buckle under the point load; this is often considered to be a failure, but is not a total collapse and would not permit a person to fall through the roof.

BS 6399 recommends that for roofs with general access the point load should be increased to 1.8kN. In the case of metal-deck roofs, foot traffic loads are applied to the membrane, and transmitted through the insulation to the deck. Weak insulation could collapse due to the point loads, if they were accidentally applied directly over a trough in the deck. For this reason it used to be customary to design metal decks with narrow troughs (75mm maximum). As insulation requirements have increased, insulation boards have become thicker; the risk of foot traffic damage has decreased, and troughs have increased in width.

When roofing materials cannot support foot traffic, or fail the 0.9kN point load test, the building must carry 'Fragile Roof''signs. This would apply to fibre cement, thin metal sheets and some plastic sheets.

Another form of roof loading results from thermal stresses. The external roofing material is exposed to direct sunlight, and its temperature can rise as high as 80°C for dark colours, or 60°C for pale colours. On clear, cold winter nights the temperature may fall as low as -15 to -20°C, so the total annual temperature range is about 100°C.

The amount of expansion or contraction as a result of changing temperature varies with different roofing materials. The length of aluminium sheets increases by 1mm in every metre for a 45°C rise in termperature. GRP rooflights expand at almost the same rate as aluminium, and steel at about half the rate. Tiles and slates are used in small pieces, so thermal movement is not a problem.

It is theoretically possible to fix roofing materials rigidly so that thermal movement cannot occur. However, this is not a practical possibility, as the restraining forces would have to be enormous. If there is no planned provision for thermal movement, it is accommodated with various effects on the roof structure, some of which are illustrated in Figure 16.5. Fixing screws, for example, may be bent by the forces, and the deformation allows the sheet to expand, thus reducing the force. The purlins may be bent or twisted, again allowing some relief of the thermal force. If the structure proves very rigid, the holes may become elongated at the fixings, and in extreme cases this could lead to roof leaks.

Thermal movement may also be accommodated by bowing of the sheets and, to a more limited extent, by stretching or compression of the sheets.

Special provision for thermal movement in aluminium sheets is usually made when the length exceeds about 10m. Figure 16.6 shows a an end lap expansion joint which allows free thermal movement while remaining weather tight. Similar provision should, theoretically, be made for steel sheets when the slope length exceeds about 15m, but this is rarely done in practice, and no major problems seem to result.

Composite panels require special treatment, because the outer and inner skins can be at very different temperatures (if the thermal insulation is working!), which can create complex patterns of thermal stress within the thickness of the panels. Again this factor is particularly important for aluminium, but should not be ignored for steel.

Secret-fix systems usually have their own unique method of accommodating thermal movement. The manufacturer's product literature will provide details; most systems employ sliding clips, rocking clips, slotted holes, or other patented devices.

Before leaving the subject of loading, it should be emphasized that the roofing materials must be strong enough to support the loads, and transmit them to the roof structure. This will usually involve the use of fixing devices, commonly screws or nails. The fixings must also be strong enough to transmit the forces.

The strength of a fixing may mean different things to different people, but perhaps

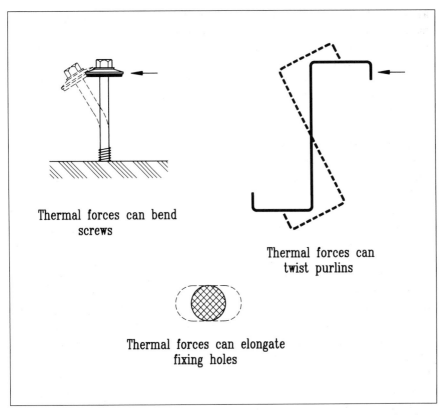

Figure 16.5   Some of the ways by which thermal movement is accommodated.

the most useful definition is 'the force at which the roofing material becomes detached from the roof '. The point here is that the fixing can fail in many different ways; each mode of failure would allow the material to become detached, but the one where this occurs at the lowest force is the critical design case.

A screw or nail may pull out of the supporting structure, (it is easy to imagine this in the case of nails into very thin timber, or screws into very thin steel). Screws and nails may also snap under tension, or fracture under shear forces. Figure 16.7 shows how screws may pull through profiled metal sheets, or cause collapse of the ribs. Some membranes may be torn around the fixings. In all of these cases the fixings have failed; the design should seek to prevent such failures by using adequate numbers of fixings, sufficiently large washers, suitably proportioned purlins, etc.

Screw fixings are often used with washers. The washer must have adequate strength and stiffness to support the design loads. If a washer is permanently deformed by the applied loadings, it will be incapable of making a secure seal.

Roof loadings must not only be supported, but there must an adequate margin of safety. There are no fixed requirements for safety factors, but it is usual to use a safety factor of 1.6 for snow loads, and 1.4 for wind loads. This acknowledges the more

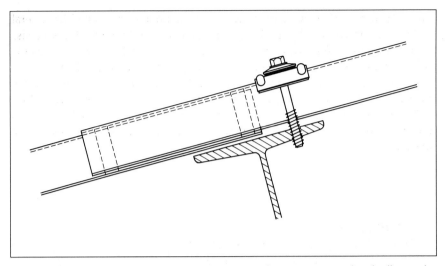

Figure 16.6   An end lap to accommodate thermal movement, and typically used for profiled aluminium.

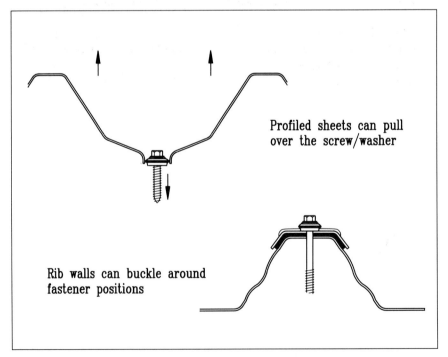

Profiled sheets can pull over the screw/washer

Rib walls can buckle around fastener positions

Figure 16.7   Possible modes of failure for fixings in profiled sheets.

permanent nature of snow loads, which can be continuously applied for several weeks, by comparison with wind loads, which are applied for only a few seconds.

The safety factor for fixings is usually taken as at least 2.0. This is a sensible precaution, as fixings are applied on site, possibly under adverse conditions, and to a variable standard of site labour. It is also possible that fixings will deteriorate due to corrosion, before the roofing materials have reached the end of their useful life.

This chapter has described loading conditions common to most roofs. Designers and specifiers must however seek to ensure that all loading conditions can be supported with safety.

STANDARDS

The Building Regulations Part A1 - Loading

BS CP3:Chap V:Part 2:1972 - Wind loads.

BS 5427:1976 - Code of practice for performance and loading criteria for profiled sheeting in building.

BS 5534:Part 2:1986 - Design charts for fixing roof slating and tiling against wind uplift.

BS 6399:Part 3:1988 - Code of practice for imposed roof loads.

FURTHER READING

BRE Digest 295 - Stability under wind load of of loose-laid external roof insulation boards.

Trade literature.

GEORGE GILMOUR
INSULATED ALUMINIUM GUTTER SYSTEM

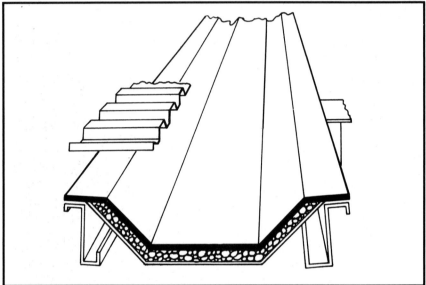

COMPOSITION & MANUFACTURE

Gutter Top Skin: Manufactured as for our long length system.

Insulation: Can be supplied in a variety of types to suit general roof specification. Normally either polyurethane or isocyanurate boards are used from standard thicknesses to achieve the required 'U' value.

Liner: The liner is manufactured from a variety of materials – generally to suit the building interior. The thickness of the liner varies to suit the support system, and is normally either painted aluminium, mill finish aluminium, galvanised steel or plastisol coated steel. During manufacture the insulation is bonded to the formed liners, which are then mechanically fastened to the long length gutters.

DESIGN

As for long length gutters a full technical backup service is provided, and installation drawings indicating laying directions, and lapping arrangements to ensure continuity of insulation are supplied.

RAINWATER PIPES

A full range of Aluminium Rainwater Pipes available in circular, square or oblong sections. Branches, Offsets, Rodding Eyes and Security Systems available.

SITEWORK

Long Length Gutters, are lightweight, and simple to handle, with the minimum amount of sitework. When isolation tape (if required) has been fitted to the steelwork, and the sealing strip placed at expansion joints, the gutters are generally craned into position.

INSPECTION, TESTING & MAINTENANCE

All gutters sections are inspected before leaving factory, and should be tested in accordance with the relevant Local Authority.

Aluminium is ideally suited to the majority of industrial, and aggressive atmospheres, and naturally forms a protective oxide layer, which is self healing if damaged. This cuts maintenance to a minimum.

Gutter outlets, and gratings should be inspected, and cleaned annually or more frequently if in an industrial environment or near trees.

George Gilmour (metals) Ltd
245 Govan Road, Glasgow G51 2SQ.
Telephone: 041-427 1264
Telex: 779210   Fax: 041-427 2205

ROOF DRAINAGE

17

Plate 17.1   10 000 metres of insulated gutters were installed at the Nissan car plant in Washington, Tyne and Wear. (By courtesy of George Gilmour Ltd.)

One of the most important functions of a roof is to prevent rainwater from entering the building. This is normally achieved by means of a properly designed and constructed weather skin, and by means of the roof pitch which causes any water to drain to the lowest point, usually the eaves.

There are several options open to the designer to accommodate the water reaching the eaves. It can be allowed to spill off, outside the building, as a cascade or waterfall; it can be channelled down the wall, for example by using curved or mitred profiled sheets at the eaves; or it can be collected in an eaves gutter and directed to the drains by way of rainwater pipes.

As these are all valid methods, each merits some attention. Allowing the water to spill off the eaves is a low-cost option, which may be attractive for some applications. However, it is best restricted to small, low-level roofs such as cycle sheds and carports. Large roofs collect great quantities of water, and this would become a real hazard if it were allowed to fall from a considerable height. It could also erode the ground on which it fell, saturate the ground around the building, blow back against the brickwork or windows, and splash dirt against the walls.

In the case of some industrial buildings, this effect is controlled by the use of curved eaves sheets (or by mitred eaves sheets, which fulfil the same function with a different appearance). The roof sheets continue to the ground, and the water is collected in purpose-made drainage channels. For such designs to be successful, the walls should be free of doors or windows or these features should be accommodated by the provision of local guttering at door and window heads.

In most other cases the water will be collected in eaves gutters, and it is to these cases to which the remainder of this chapter is directed. Sometimes stormwater is allowed to flow around a curved eaves sheet before being collected in a high-level gutter, but for the purposes of this chapter, these gutters will also be regarded as eaves gutters.

Perhaps the first consideration in any drainage design process is to establish the amount of water to be collected, channelled and drained. In Britain this process is simplified by the existence of a comprehensive British Standard BS 6367. The driest areas in Britain are in the south east of England, around Kent and East Anglia, and the wettest regions are in the west, particularly in Scotland and North Wales. It would seem reasonable to assume that those regions which receive the highest annual rainfall will require the largest and most secure gutters.

In fact, this is an incorrect assumption! The design condition for the drainage system is not governed by the amount of water which passes through it in a year, but by the most extreme conditions it is likely to experience. It is easy to imagine a system of gutters and rainwater pipes coping quite adequately with continuous, gentle rainfall. However, it is obvious that, if the rainfall were to be steadily increased in intensity, there would come a time when the system was overloaded and the gutter would overflow.

BS 6367 shows that the regions which suffer the most intense storms are not the same as those which have high annual rainfall. It has been shown that a two-minute storm is sufficient to flood inadequate systems. The standard identifies the areas where the heaviest storms occur; a simplified map, showing these, is given in Figure 17.1. This shows that the regions suffering the most severe rainstorms almost never correspond to the regions suffering the greatest annual rainfall.

The standard should be used to identify the design conditions for a building. It offers guidance on the probabilities of storms of exceptional severity within the life of the building. It also helps in establishing suitable factors of safety when higher-than-average security is required.

There is no single recommended shape for gutters, but there are several very popular shapes which, with minor variations, will be encountered with great regularity. The most common of these are shown in Figure 17.2. *Half round* gutters are extensively used on housing, and have a major share of the domestic market. Typical sizes are 75 and 100mm diameter, but 150mm diameter is sometimes used on industrial buildings. In the past, gutters were often made in cast iron, but these have now been superseded by PVC, which is light in weight, easy to cut, and virtually

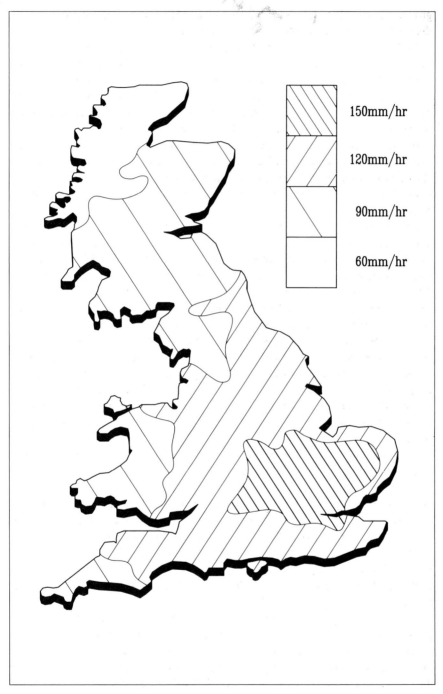

150mm/hr

120mm/hr

90mm/hr

60mm/hr

Figure 17.1   Approximate variation of rainfall intensity, based on
maximum sustained rate for two minutes, once in five years.

maintenance-free. The larger industrial half-round gutters are sometimes formed in fibre cement.

Decorative forms of *rectangular box* gutters are sometimes preferred to half round for aesthetic effect. For domestic applications these are available in PVC, and in extruded aluminium with plain, coated or anodized finish. These would be up to about 100mm wide and 75mm deep.

Larger rectangular box gutters are often used on factories and warehouses. These are, typically, 150 to 200mm wide and 100 to 150mm deep. The most common material for these is galvanized mild steel, with a substantial coating. Other, more durable possibilities include aluminium, fibre cement and GRP.

Boundary wall gutters originally had a special significance which has now largely disappeared. If the owner of a piece of land wishes to make the greatest possible use of it, as a building site, he will build a wall on the boundary. However, he may not have any part of the building outside that wall as it would then be outside the limit of his site. He must therefore locate the gutter immediately inside the wall line. Today it is quite common for boundary wall gutters to be chosen on aesthetic grounds, even when the wall is not close to the site boundary.

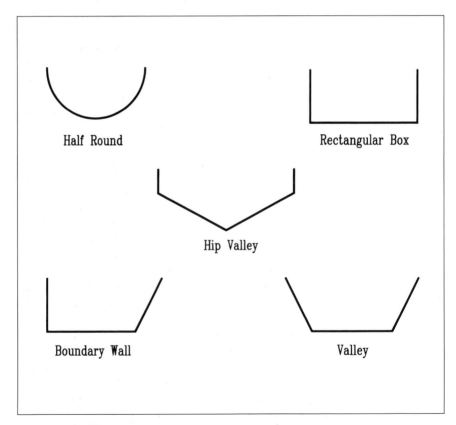

Figure 17.2 The most common gutter cross-sections.

This point merits further consideration. The roof has performed its function by intercepting the rainfall and directing it to the eaves. It is important that this water should be denied access to the building, otherwise the roof contribution is wasted. It is therefore logical to locate the gutters outside the building where a gutter failure is a minor nuisance rather than a major catastrophe. A true boundary wall gutter cannot be replaced with an external gutter, but an aesthetic one could; the designer should not choose to increase the occupier's risks unless he has very strong reasons for doing so. (Of course, a boundary wall gutter will always be required behind a parapet.)

A *valley gutter* is located between two roof slopes, i.e. between successive roofs in a multi-span arrangement. This is a case where the gutter cannot be external to the building. If the gutter fails, water will enter the building. It is therefore sensible to design valley gutters with increased factors of safety, and to select the most durable materials.

Valley and boundary wall gutters are available as standard items in fibre cement, but are usually manufactured to order from aluminium, coated galvanized steel and GRP. It is quite common for large gutters to be used as walkways; where this is intended, the materials may need to be increased in thickness to ensure sufficiently robust construction. Mild steel gutters are often about 3mm thick in order to prevent buckling during galvanizing. Aluminium gutters can be 2mm thick for widths up to about 150mm, and 3mm for widths up to about 300mm.

Another form of valley gutter is required at hip valleys. These are not usually large, as they drain quickly due to their pitch. In tiled and slated roofs these gutters are often formed on site, from lead or high-purity, soft-temper aluminium. In industrial roofing applications, they are usually purpose-made in aluminium or coated galvanized steel. These gutters do not have flat soles; the sole is in two halves, with one half in each roof slope. The shape geometry requires careful calculation.

Any gutter which is located over the building (e.g. valley and boundary wall) will usually require insulation. An uninsulated gutter would suffer large amounts of condensation on the underside, and would allow heat to escape from the building. However, if the insulation is efficient there is a greatly increased risk of water freezing within the gutter; the worst scenario would be a sudden thaw with snow and ice on the roof melting before that in the gutter. This would lead to an overflow with water pouring into the building. The usual recommendation is to provide about half the insulation under the gutter compared with the remainder of the roof; this should ensure that there is free drainage in the gutters before the main thaw occurs.

If the gutters are to be used as walkways, the insulation must be sufficiently rigid to support the foot traffic without becoming compressed. It will usually be of board or rigid slabs.

In felted or single-membrane roofs, the gutter can be formed as a channel at the edge, or between slopes. Again there can be reduced insulation at these positions.

Some designers simply locate outlets at the lowest point in the roof, without creating any defined channels. The argument in favour of this is that the roof is fully waterproof, so there can be no overflow. In extreme conditions, the level of water will rise, with the roof acting as a reservoir; later the water will drain when conditions permit. There is a serious risk here, however, because water is heavy and could overload the roof; if the depth of water exceeds 150mm it is likely to exceed the design allowance for snow loads. If outlets become blocked, the water level could continue to increase until collapse occurred. In any case, standing water should be avoided, as this carries an increased risk of roof leaks.

All materials expand when heated, and contract when cooled. The effect is more pronounced in aluminium and GRP than in steel, but is always present to some extent. Gutters can be applied in considerable lengths, especially when successive lengths are rigidly connected; it is often necessary to make provision for thermal movement.

In the case of both aluminium and GRP, an allowance of 1mm for every metre of gutter length is usually sufficient. Half of this allowance should be sufficient for steel gutters. Rainwater pipes are normally fixed to the building structure and are therefore fixed relative to the gutter. It is best to incorporate an expansion joint between successive pipes to prevent the build-up of thermal stresses which could damage the pipes or their fixings.

Figure 17.3 shows some popular arrangements for expansion joints. The simplest form consists of two stop ends back to back, with a space of 25–50mm between; a cover flashing is fixed at one side only. This is inexpensive, and is easy to make. However, it has certain drawbacks; if there is only one rainwater pipe between

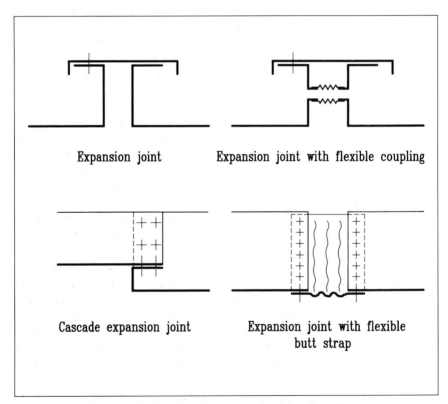

Figure 17.3   Some methods of forming expansion joints in gutters.

successive expansion joints, then a pipe blockage must result in an overflow. It follows that this detail is only suitable for external gutters.

The arrangement can be improved by the use of a flexible pipe coupling between the stop ends. Movement can still occur, but there is now some provision for overflow from one gutter to the next; this is vitally important if a rainwater pipe becomes blocked.

For large internal gutters, a cascade-type joint is sometimes used. The joint is formed with slotted holes to allow movement, but it is constructed in such a way that leaks are contained within the gutter. These have the advantage that the gutter size becomes greater as the volume of water collected increases.

Some traditional tiled or slated roofs have a dwarf parapet for aesthetic effect. Such roofs often employ lead gutters which are formed *in situ*. The lead cannot be applied in very great lengths because of potential problems of thermal movement. It is therefore normal to use the stepped arrangement, with the gutter becoming deeper and narrower as it approaches the outlets.

It is also possible to use to use flexible joint straps between the lengths of gutter. These are usually made from neoprene or EPDM, which are extremely flexible, but are eventually degraded by ultraviolet radiation. Such details demand regular inspection, and may require replacement as part of a maintenance schedule.

Specialist gutter manufacturers have their own special details, and are usually pleased to advise on specific applications.

It is quite common to encounter arguments as to whether gutters should be level or inclined. BS 6367 bases its design methods on the assumption that the gutter is level. For domestic applications, using shallow half-round gutters, it is good practice to incorporate a slight fall; this is really to ensure that settlement does not create negative falls, and a fall of 1 in 350 is suitable.

A house that is only 6m long will require a gutter fall of about 18mm, so the gutter level will not fall too far below the eaves line. By comparison, a factory may be 200m long; even a 1 in 350 slope would result in a fall of over 0.5m. Large gutters do not need to be laid to falls, as water creates its own gradient between rainwater pipes. The water surface is low at an outlet, and high midway between outlets, and this produces an effective hydraulic gradient which causes the water to flow towards the outlets. Deep gutters can create greater effective hydraulic gradients, and can therefore achieve greater rates of flow; a small increase in depth can be more effective than a large increase in width.

Gutters are usually supported on purpose-made brackets, which are spaced at centres of up to 1m. Good gutter brackets have some means of adjustment, so that they can be levelled. This is an important feature, as structures can settle or be built slightly out of level, and some compensation is needed for accurate gutter alignment.

In domestic applications, the brackets are usually screw-fixed to a fascia board; they are set to a string line to follow a gentle fall. In industrial buildings, the brackets are supported on the eaves purlins, or eaves beams; packs or shims are used to maintain the level.

Not only does BS 6367 provide information on intensity of rainfall, it also provides guidance on establishing the size of gutters and the spacing of rainwater pipes. A rainfall intensity of at least 75mm per hour should be used, but this may be increased for greater security. 75mm per hour represents 0.0208 litres per second on every square metre of roof.

The standard gives formulae for calculating the capacity of gutters. These must be adjusted to take account of bends (for example, around bays or recesses), or friction in very long gutters. The gutter depth is taken as the actual depth for external gutters, or a reduced figure for internal gutters. This is to provide security against water splashing over the sides of an internal gutter. The method of calculation assumes that the gutter is level, and the standard proposes that any fall should be viewed as an additional margin of safety.

Figure 17.4 shows the flows which are possible in simple rectangular box gutters of various widths and depths. The calculation is based on the method in the standard; it is assumed that the gutters are level, drain freely, are straight, are relatively short and are for internal applications. Valley or boundary wall gutters of similar sole width will have slightly greater capacity. The figure is intended to provide a useful design guide, and gives a good indication of required gutter size for a given application. It is *not* a substitute for the British Standard which should always be used for accurate design. A designer who regularly needs gutter design calculations should prepare his own data sheets along the lines of this example.

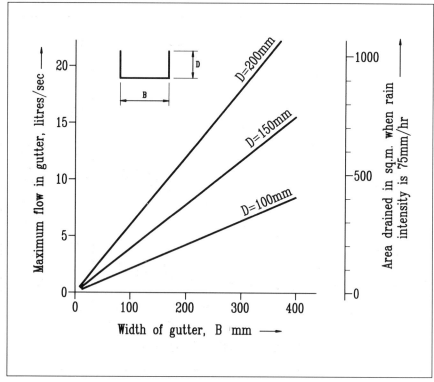

Figure 17.4   Approximate drainage rates through rectangular gutters.

The standard has provided the means of calculating the maximum amount of water arriving at the gutter, and the maximum amount which can flow through the gutter. However, this is only useful if the water can escape from the gutter at the same rate. Again, the standard provides formulae to calculate the maximum discharge rates through pipes of different diameter. It is shown that the amount increases as the head of water over the outlet increases.

Figure 17.5 was prepared from the methods in the standard. Simple circular outlets were assumed, of the same diameter as the pipe. Again this is intended to provide a design guide, and use of the standard is essential in making accurate calculations. This is another situation where a regular user may wish to prepare data sheets to cover his specific applications.

Reference to Figures 17.4 and 17.5 in conjunction shows that a gutter can only achieve maximum flow rates if there is a reasonable head of water over the outlets. However, this is the very place where the water level is lowest, and the head of water required may not be a practical requirement (remember that the upstream depth must be greater in order to create a hydraulic gradient).

For small buildings having a roof area of up to about 100 m² and with half-round eaves gutters, it is possible to adopt a much simpler approach to gutter design. Part H3 of the Building Regulations provides simple tables from which the sizes of rainwater pipes and half-round gutters may readily be obtained.

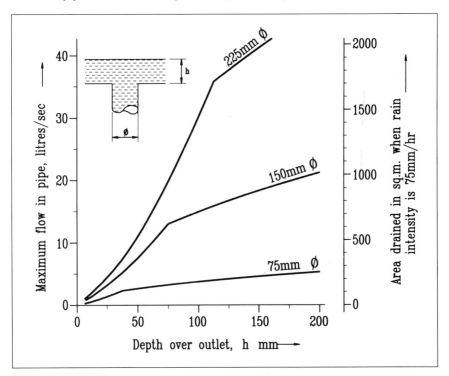

Figure 17.5   Approximate drainage rates through circular rainwater pipes.

Specialist manufacturers of gutters may also provide design guides to assist the user to establish gutter sizes, pipe diameter, outlet spacing, etc. Reference to a graph or table is usually much less time-consuming than carrying out laborious calculations.

Figure 17.6 compares some of the common outlet details. The simple outlet is economical, but may not provide sufficient head of water for maximum drainage. The sump arrangement is a means of producing a greater head of water, but this is at the expense of complication; the structure must leave clear space for the sump, even if this entails notching of purlins or eaves beams.

It is the diameter of the actual outlet which governs the maximum rate of flow, so using conical outlets can be as successful as increasing the pipe diameter. In fact, the outlet diameter can be 50% greater than the pipe diameter, and provided the transition is over a reasonable distance, the drainage rate can be based on the outlet rather than the pipe. In this context, a reasonable distance is a length at least equal to the pipe diameter.

A prudent designer will always envisage the possibility of blockages in drainage systems. This can easily result from an accumulation of leaves or straw, or from a dead bird. It is sensible to include some form of grating over the outlets; this will prevent major blockages, but the gratings must be cleared as part of a regular

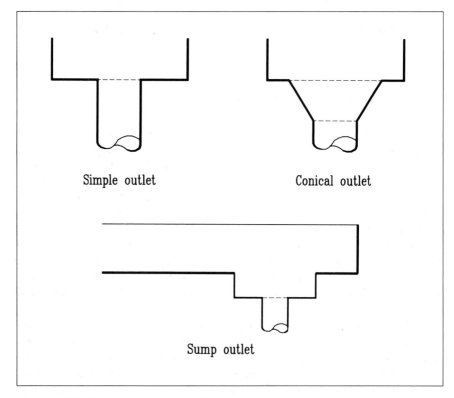

Simple outlet   Conical outlet

Sump outlet

Figure 17.6   The most common outlet arrangements.

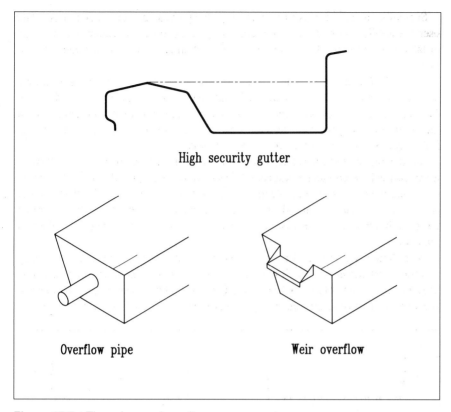

High security gutter

Overflow pipe   Weir overflow

Figure 17.7 Three forms of overflow arrangement.

maintenance routine. It is also possible to introduce details which provide warnings when blockages have occurred. Some typical examples are illustrated in Figure 17.7.

The high-security gutter provides two different forms of security. Because its inner wall is taller than its outer wall, any overflow will be outside the building. Such an overflow is clear evidence of a blockage. It also gives security in that its outer edge will not provide purchase for a grappling iron; it is therefore popular with designers of prisons and other high-security buildings.

The overflow pipes and weir overflows can project through walls or parapets. It must be emphasized that these are intended to provide a warning that the rainwater pipes are blocked, but they do not provide an alternative drainage system. In an industrial application the rainwater pipes are usually 150mm diameter or more, and cannot be replaced by a 50mm pipe through the gutter wall!

Building owners should be encouraged to arrange a regular maintenance schedule. The work involved in clearing debris from gutters is very small, and normally is only required annually. By comparison, the damage and disruption which can result from a gutter overflow can constitute a catastrophic event. A relatively small amount of rainwater can cause thousands of pounds worth of damage to stored materials or electronic equipment.

There is a vast range of proprietary systems of gutters and rainwater pipes. Manufacturers provide design data for their products, and recommend suitable details.

This chapter has concentrated on the traditional, conventional systems, but other novel or innovatory systems may be encountered and these include syphonic systems and pumped drainage.

Syphonic systems are designed to create a partial vacuum which pulls water through the pipes. One of the great benefits of such systems is that the rate of flow in the pipes is much greater than for simple gravity systems; this reduces the number of pipes required, and reduces the risk of blockage. The main disadvantage is that these systems are complicated and should be designed and installed by experts.

It is also possible to create drainage systems which rely on the use of electric *water pumps* to transfer large quantities of water to convenient drains. The advantage of such an arrangement is that the drains may be kept outside the building; the great disadvantage is that a power failure can result in flooding. Again, such systems should be left to experts.

A roof which has been properly designed and constructed will collect the rainwater and distribute it to the eaves of the building. Storm water which has been brought under control should be kept under control, and this is most easily accomplished by means of a suitable system of gutters and downpipes.

STANDARDS
BS 460:1964 - Specification for cast iron rainwater goods.

BS 569:1973 - Specification for asbestos-cement rainwater goods.

BS 1091:1963 - Specification for pressed steel gutters, rainwater pipes, fittings and accessories.

BS 2997:1958 - Specification for aluminium rainwater goods.

BS 4576:Part 1:1989 - PVC half-round gutters and pipes of circular cross-section.

BS 6367:1983 - Code of practice for drainage of roofs and paved areas.

FURTHER READING
BRE DAS 55 - Roofs: eaves, gutters and downpipes - specification.

Trade literature.

SITE OPERATIONS

18

Plate 18.1   Ropes and harnesses are the most practical safety devices when roofing a dome. (By courtesy of Alcan Building Products.)

When a roofing project is planned or contemplated, the important decisions are usually made in offices remote from the site.

The architect will probably decide on the appearance and performance requirements and he may discuss these with manufacturers of roofing products. An engineer will design the roof structure, and the local authority will be involved in approving the design. General contractors submit tenders and, subject to the agreement of the building owner, one of these will be appointed. The general contractor may employ a roofing subcontractor who will prepare the detail drawings, and agree a programme for the work.

For the project to succeed the design must be correct, the materials must be of suitable quality, and the work schedule must be appropriate to the type of construction involved. The standard of workmanship is also of crucial importance, of course: concepts developed and specified in the safe, comfortable surroundings of a design office, must be translated into reality on an exposed, inclined roof slope, in whatever weather conditions prevail, so that the skill of the tradesman must match the ingenuity of the designer.

It is not the purpose of this chapter to teach roofing skills; these can only be acquired by practice, under the guidance of an experienced superviser. However, the designer cannot be expected to produce first-class solutions without a reasonable appreciation of site conditions and practices. The following pages are intended to assist in providing this appreciation.

Roofing operations are at the mercy of the weather. The work cannot be sheltered, and the site programme may depend on the roof being completed early to provide shelter for other trades. Some operations are impossible in wet conditions, and the programme should take account of the risk of lost time.

There is usually a breeze at roof level, even when conditions appear to be calm on the ground (and a wind at roof level when there is only a breeze on the ground!). This can create difficulties in handling large pieces of lightweight insulation, profiled sheets, felts or membranes. When a site is known to be exposed, the designer should consider the use of smaller units of material, or allow for more lost time due to inclement weather.

The designer must also be reasonable in his demands for accuracy. It is very easy to show a detail on a drawing requiring site cutting to the nearest millimetre. The tradesman on the roof has no workbench or vice; he will carry out the cutting while kneeling on an inclined plane (perhaps on a roof ladder), with the wind threatening to blow away his work piece. The designer should try to produce details which provide a degree of tolerance.

Roofs are dangerous places to work. Their exposed position means that they can be slippery from moisture or frost, they are sloping, and are not always made of materials that can support the weight of a man. The roofing industry has experienced many tragic accidents in which people have fallen off roofs, or through part-finished roofs. As a result of these accidents, many improvements have been made in site safety techniques, but constant care is essential in roofing work.

There are various legal requirements which affect work on building sites, and one of the most important is the Health and Safety at Work Act of 1974. This places a duty on employers to provide a safe place of work, and a safe system of working. The employer must ensure that the worker understands the method of working, and the reasons for its adoption.

In practice, this often requires that scaffolding is erected to meet specified standards, that guard rails are fitted around roofs under construction, that roof ladders and stagings are supplied, and that protective clothing is issued.

The Health and Safety Executive (HSE) provides guidance on site safety matters, and has powers of enforcement. There are HSE leaflets and booklets on a wide range of roofing topics. It is required that guard rails are fitted around any work area from which a worker could fall more than 2m. Similarly, a barrier must be fixed whenever there is a danger of materials falling from the roof. (Note that barriers are required around rooflight openings as well as around the perimeter of the roof). The means of access to the roof must also be safe and secure, with rails or barriers as a protection against falling.

For very small projects, it may be sufficient to use ropes and harnesses, but some protection against falling must be provided. When a job is likely to last more than six weeks, the local office of the HSE must be informed, so that inspection visits may be arranged.

Throwing waste materials fom a roof is dangerous, and is also illegal! Some form of rubbish chute must be provided. This ensures that all waste will fall to the correct place and cannot be whisked away by a sudden gust. When hoisting material, it is important to ensure that the public is prohibited from entering the work area, and that safety hooks are used.

There are further rules and suggestions for those occasions when it is necessary to use a bitumen boiler. The boiler should be set up on a level surface, spare gas cylinders should be kept at least 3m away, the maker's instructions should be followed, and a fire extinguisher must be kept close by.

Workers should also be supplied with all essential items of safety clothing, such as helmets, gloves and goggles.

On some projects, the workers are issued with harnesses which can be tethered to the building structure. These provide complete security against falling, but workers are often reluctant to wear them on the grounds that they impede free movement, slow down the work, and reduce productivity bonuses. Supervisers must be very firm in ensuring that all safety equipment is used, and designers must be aware of working methods in order to make practicable designs.

Perversely, the roofing professional may not always enjoy the same safety standards as the roofing tradesman. An architect, surveyor or roofing contractor may be called to inspect a roof, with a view to possible renovation, repair or extension. He is likely to be offered access, by means of an extending ladder, to a roof lacking any form of safety rails or protective barrier. There is no shame or disgrace in refusing to proceed until adequate safety standards are established.

Proper work practices ensure the safety of the tradesmen, but it is their skill which has a major influence on the success of the roof. This skill is acquired by training and experience. Experience is gained by working as a member of a team; the leader allocates tasks, taking account of the ability of the various team members. Thus the junior member of the team gradually learns new skills and techniques; in due course he can move up as a new junior member is introduced. This traditional approach can only be successful if the team leader is experienced in the type of roofing being undertaken. In recent years there have been rapid changes in roofing design and roofing materials, and the training of team leaders has become an essential requirement.

This can be illustrated by the example of industrial roofing. Consider the case of a youth leaving school around 1918 and starting work as an industrial roofer. The skills he acquired in his first few days were to do with single-skin asbestos cement. He was taught to cut mitred corners at overlaps, to use a 'fiddling stick' to fit hookbolts, to protect the exposed bolts, to seal lapped joints, to carry sheets up ladders, and so on.

The youth in our example would have seen very little change in industrial roofing during his entire working lifetime. The lessons he learned on his first day at work were equally valid on the last day before his retirement. He could have passed on these lessons to numerous apprentices and trainees.

A second youth, starting 20 or 30 years later (perhaps the son of the first youth!) would have had very different experience. For 20 years, single-skin asbestos cement was the almost universal material for industrial roofs; then came a bewildering succession of rapid changes.

Single-skin gave way to double-skin with a thin layer of insulation quilt. The insulation thickness was increased, and timber spacers were needed to prevent its becoming compressed. The inner skin of asbestos cement was replaced by plaster-board, then this was replaced by profiled metal. The spacers were changed from timber to galvanized steel. Changes of insulation were tried, particularly poly-urethane board and expanded polystyrene. Composite panels were introduced, in both steel and aluminium; these were produced by injecting foam insulation, or by bonding insulation boards to profiled sheets.

The widespread use of thicker insulation gave rise to condensation problems, and these were solved by the introduction of vapour checks, breather membranes and ventilated filler blocks.

Fasteners changed from hookbolts in the profile crown to self-tapping screws in the crown, then to self-tapping screws in the trough. Next, self-drilling screws became available, then the fastener material changed from zinc-plated carbon steel to stainless steel. Side laps were closed with POP rivets, then blind rivets, then expanding blind rivets, then self-tapping screws, and finally self-drilling screws.

All of these changes occurred within about 20 years, and many were simultaneous. The roofing fixer on the building site was no longer repeating the same operations he had been taught as an apprentice. Almost every job included new features, and some products demanded a change in established practice (e.g. drill speeds must be varied to suit the type of screw).

To keep pace with these developments, regular training became necessary. Some of the larger companies set up in-house training courses for their staff; some manufacturers provided training in the use of their specialist products.

Unfortunately in the roofing industry there are many small teams which operate as self-employed subcontractors. These teams may not have had appropriate train-ing; they may also be employed on the basis of a fixed-price arrangement, which gives every incentive to complete the work as quickly as possible, but no real pressure to do it well. When special skills are required, the employer should demand proof that the employee has received proper training, and that he has understood it.

There are some skills which are so specialized that it is always advisable to sublet them to an expert. These include site-welding of aluminium, and site-painting of steel roof sheets. Again, it is important to ensure that the 'expert' really has the skills required.

There are great pressures to increase productivity in roofing operations. Because work can be disrupted by inclement weather, it is essential to make the greatest possible progress whenever the weather is fine. Also, roofing is just as competitive as any other building work; the lowest tender is likely to win the contract and, as the material costs are the same for everyone, reducing fixing costs offers the best chance of producing the lowest tender.

There is a steady stream of new ideas intended to increase productivity: new types of fasteners (e.g. self-drilling screws), new types of tools, and new products such as composite panels, which allow the lining, insulation and outer sheet to be fixed in a single operation. Screws can be supplied in the form of continuous belts which are fed into automatic screw guns.

The roof designer should always ensure that new products do not bring greater speed at the expense of quality. For example, the traditional method of cutting metal sheets was with a hand saw. Later the nibbler became popular, and this has largely been superseded by the powered disc cutter, but this should not be used on organic coated steel sheets, as its heat melts the coating and exposes bare metal. A really innovative new product should combine higher productivity with improved quality.

There are also many products which are designed to remove the need for traditional specialist skills. Examples can be found in every type of roofing.

The use of rigid, profiled underlays can simplify the task of tiling and slating. It becomes virtually impossible for the roof to leak, and removes the need for the tiler

Plate 18.2   Site-welding flashings around a large opening in a secret-fix aluminium roof. (By courtesy of Melvyn Rowberry Welding Services Ltd.)

to fit felt underlay with the right degree of sag, or to fit counter battens. Pressed metal panels to simulate roof tiles remove the need for the tiler's skills altogether.

When fixing profiled sheets, self-drilling screws exclude the possibility of drilling pilot holes to the wrong diameter; pre-formed sealer strips are the correct size, and do not require judgement on the part of the user. Secret-fix systems can produce a weathertight roof, even when the fixer lacks the skill to tighten the screws correctly (and pre-set screw guns can tighten screws correctly, even in inexperienced hands). Composite panels obviate the need for skills in jointing vapour barriers, or fitting insulation.

It is possible to purchase roof panels consisting of metal bonded to plywood or chipboard; the metal has part-formed seams at the edges. These can simplify the task of laying continuously supported metal, but are probably less versatile than the traditional method.

Felts are available in 'torching grades' which have a backing of bitumen. They are heated on site by means of a butane torch, which melts the bitumen as the felt is rolled out on the roof. The felt is simply rolled into place, and the bitumen then cools and sets. This removes the need to heat bitumen on site, and dispenses with the skills involved in spreading the bitumen evenly and in the correct quantity. However, some skill with the butane torch is necessary, the bitumen must be heated sufficiently to ensure an even and continuous bond.

The building industry, as a whole, suffers high levels of material waste; roofing is no exception to this. Waste can occur in many ways. Materials can be damaged by clumsy handling, by unsuitable storage, through theft, through poor site control, by incorrect application, and by abuse from following trades.

Theft can be largely eliminated by the provision of a secure storage compound. A screen or fence is a deterrent against pilferers, it stops unauthorised personnel wandering about amongst the stored products (and perhaps inadvertently causing damage), and it should prevent vehicles from driving over the products. Small, valuable items and tools should be locked inside the site hut when not in use.

Materials should be covered whenever necessary by tarpaulins or plastic sheets. Insulation boards and quilts could become saturated if inadequately protected, and lining sheets could become dirty. Even stacks of metal sheets are stained if water enters between the sheets in the stack — and the water could be condensation as well as rain. Profiled sheets should be stacked clear of the ground, on baulks of timber, and on a gentle slope to shed moisture.

On some sites, the ground seems to be covered by screws, rivets, shaped washers, extruded clips, plastic colour caps, and so on. These are expensive items and the waste represents a lot of money; better site control and supervision could prevent much of this waste.

The materials should be ordered in the correct size for the job, and the fixers should ensure that they are using the correct pieces in each position. It is extremely wasteful to cut down a board which appears to be oversize, only to find that it was intended for a different part of the roof. Having cut up the long pieces, there are only short ones left on site!

When an area of roof is complete, and if other trades are to work over it (bricklayers, glaziers, etc.) it is advisable to request a written statement from the clerk of works confirming that the work is free from dents, cracks, or other damage. Roofers are able to walk on most types of roofing without causing damage, but other workers may be less careful, or less aware of the risks. Once the damage has occurred, everyone will deny liability, so it is wise to have the acceptance note from the clerk of works.

Site handling methods vary with the type of materials and the type of site. When a house roof is being tiled, the tiles are carried up the ladders by the tiler or his labourer; this is physically demanding work. When long profiled sheets are fitted on a large factory roof, a crane is usually used to lift these from the delivery vehicle and on to the roof. The sheets should be ordered in packs to roof a specific area, and this will avoid the need to handle them any more than necessary.

Crane hire can be expensive. It is important that the delivery of the material coincides with the time that the crane is available. Also, cranes are governed by reach as well as by load. A crane which can lift a massive weight at short reach may have very little lifting capacity at long reach. Cranes which lift materials on to roofs are usually required to cope with relatively long reach.

When lifting very long sheets (secret-fix systems can be delivered in sheets over 30m long), it is not enough to use a single or double sling in the centre of the packs; the ends of the sheets would droop under their own weight, and could buckle. A spreader beam is used; this is a rigid beam from which further slings can be suspended to support the sheets more uniformly. The weight of the spreader beam must be taken into account when ordering the crane as it is quite common for the spreader beam to weigh more than the packs of sheets being lifted.

The final piece of advice in this book on roofing refers to something which should in fact occur before the roofing work commences. This may seem perverse, but the advice is so valuable that it bears repeating.

CHECK THE ROOF STRUCTURE

Roof slopes are exposed, have restricted access, and can be dangerous. It is not sensible to compound the difficulties by attempting to work on inaccurate or unsuitable structures. Furthermore, it is impossible to construct a first-class roof if there are underlying structural faults. If the trusses do not lie in the same plane, the tiling battens will wander and the tiles will follow the battens. If the purlin flanges are not aligned, the profiled metal sheets will rise and dip to follow the purlins. If the rafters sag, the felt roof may pond over the deflected structure.

All of these are examples of poor roofing, for which the roofing contractor should not be held responsible. The roof structure may well be the result of the accumulated tolerances of all other trades. The roofing contractor should make a level and dimensional check before commencing work. Any discrepancies should be brought to the attention of the site manager or the clerk of works.

At this stage the problems can be addressed and rectified, or they can be acknowledged, and the roofing contractor instructed to proceed. In the former case a first-class roof is possible, in the latter it will be understood that the final construction cannot be perfect, and less critical tests will be applied.

The roofing contractor should not feel embarrassed at pointing out inferior workmanship to the general contractor. Rather, it is the general contractor who should be embarrassed at having his failings discovered by others.

STANDARDS
BS 1129:1990 - Specification for portable timber ladders, steps, trestles and lightweight staging.
BS 5973:1990 - Code of practice for access and working scaffolds and general scaffold structures in steel.

FURTHER READING
Roofing and cladding in windy conditions - NFRC
HSE Safety in roofwork - HMSO
HSE Information sheets.
Timber scaffold boards - reducing the incidence of site injury - BRE

INDEX

thermal conductivity 100-102, 110

thermal insulation 6-8, 34, 61, 77, 82, 86, 89, 91, 99-113, 118, 123, 130, 131, 134, 172, 209, 221

thermal movement 47, 49, 51, 62-8, 77, 81, 83, 91, 108, 133, 142, 146, 156, 168, 200, 201, 210

thermal pumping 63, 71

thermal reistance 101, 102

thermal transmittance (*U* value) 102-7, 112, 129, 130, 142

tiles 1-3, 5, 29-39, 43, 85, 91, 143, 150-3, 161, 165, 176

tiling battens 31-4, 36, 37, 40, 41, 92, 151, 152

tolerances 151, 221

torch-on 43, 82

toughened glass 135

translucent sheets 23

trapezoidal profiles 45, 46

U

U value (*see* thermal transmittance)

undercloak 151

underlay 32-4, 37, 64, 82

under-purlin lining 88, 89

V

valleys 20, 36, 39, 76, 150, 166, 194, 208, 209, 212

vandalism 18, 137

vapour barrier 7, 8, 79, 85, 94, 95, 101, 120, 131, 149, 222

vapour check 79, 95, 120, 131, 142, 149, 165, 220

vapour pressure 120, 122, 123

vapour resistance 120-2

ventilated profile fillers 123, 124, 166, 220

ventilators 24, 25, 168, 188

vent pipes 141, 168

verge clips 151, 153, 161

verges 16, 36-9, 84, 151-5, 157

vermiculite 111

W

walkways 25

warm roofs 82, 108, 109

welding 144, 145, 220, 221

welted joints 66-8, 70, 148

welts 154, 155

wind 9, 30, 32, 38, 62, 77, 78, 84, 91, 108, 121, 151, 156, 162, 191, 192, 194-8, 201, 203

wind pumping 63, 71

windows 15

wood cored rolls 62, 66

woodwool 78, 79, 91, 93, 94, 121, 180

Z

zinc 4, 50, 60-2, 69-71